InspireScience

Science Handbook

Grade 3

mheducation.com/prek-12

Send all inquiries to:
McGraw-Hill Education
8787 Orion Place
Columbus, OH 43240

ISBN: 978-0-07-679235-1
MHID: 0-07-679235-8

Printed in the United States of America.

1 2 3 4 5 6 7 8 9 DOR 21 20 19 18 17 16

Table of Contents

OVERVIEW

Inspire Science Handbook is a book all about science. The Handbook has information on many science topics. You can use this book to look up something that you want to know. For example, you can use it to learn about different living things and their ecosystems. You can also use this book to learn how to act like a scientist. For example, this book will help you understand how to read different weather instruments.

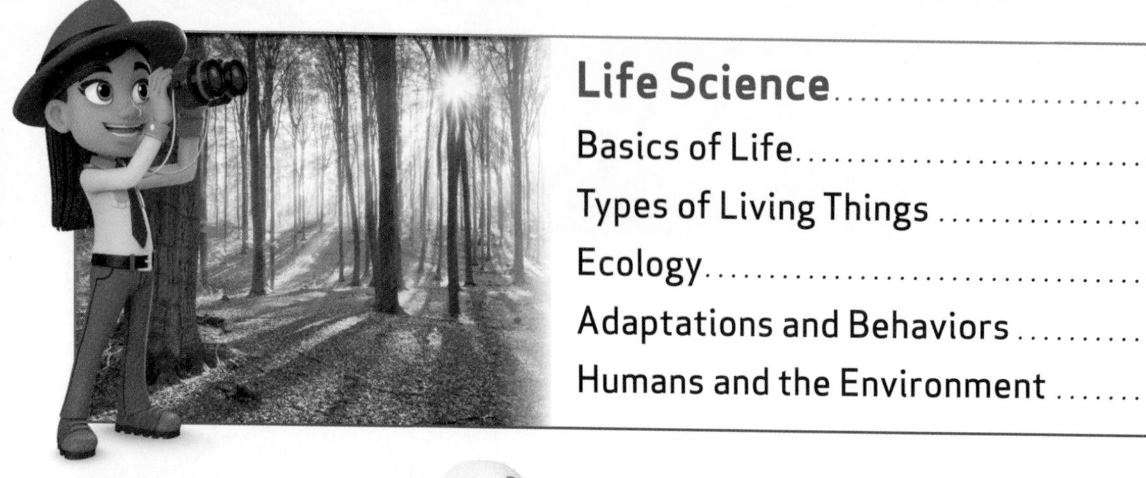

How to Use This Book

Looking at the Front Pages

At the front of the book you will find the Table of Contents. This section of the book will help you quickly locate information that you are looking for. For example, if you are looking for information about how energy is created, you would look in the Physical Science section of the Table of Contents.

Sections

Throughout the book, you will see many different titles and headings. The main focus of each section is called out in large text. Each main focus then has smaller sections with more detail about the topic. Each section also uses a different color: Life Science is green, Earth and Space Science is orange, and Physical Science is purple.

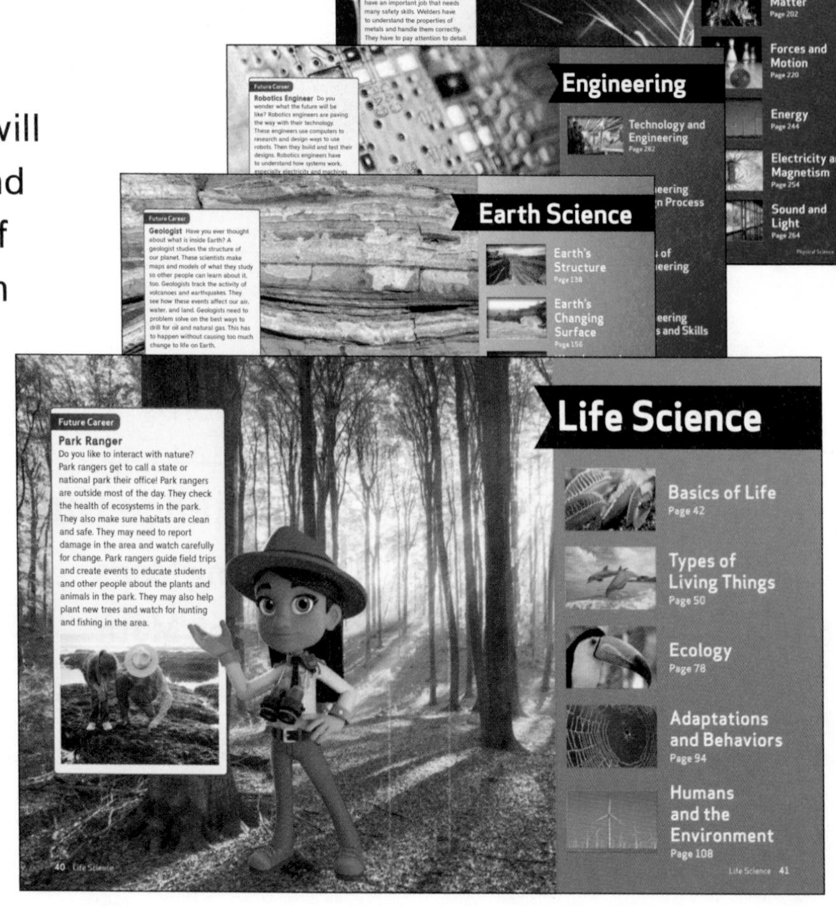

Page Features

What Could I Be? In this box, you will find careers that relate to the content on the page. Flip to the Careers section in the back of the book to learn even more!

What Could I Be? **Bacteriologist**

Interested in becoming an expert on single-celled living things? A bacteriologist studies effects of these tiny organisms on plants and animals.

The **Make Connections** box will tell you about other sections of the book that relate to what you are currently reading. Turn to the page number listed to learn more about that topic.

Make Connections

Jump to the *Human Body* section to learn about your organ systems.

The **Fact Checker** box will correct common science misconceptions. By knowing the correct science you can build on your learning and expand your knowledge about many science topics.

Fact Checker

Some kinds of bacteria make us sick, but others are helpful. They help your body break down the food you eat.

Did You Know? In this box, you will learn fun science facts that will expand your learning! You might also find ideas to research or plan your own investigations.

Did You Know?

The human body contains many levers. Your head and neck act as a first-class lever. Your foot acts as a second-class lever. Your forearm acts as a third-class lever.

The **Word Study** box will break words down to their roots, making it easier for you to understand their meaning.

Word Study

Bio comes from the Greek word for *life* as in *biology*.

In the **Skill Builder** box, you will learn how to read diagrams, charts, graphs, tables, and other graphics that can help you understand the concepts you are learning.

Skill Builder **Read a Diagram**

Notice how the part of the Moon we can see from Earth changes. The lit area visible to us grows larger, and then smaller.

Career Kids

Throughout the book you will see the Career Kids. They will help you learn about different jobs that grown-ups have that relate to specific science topics. For example, these kids can help you understand what it would be like to be a park ranger. Make sure to check out the Career Kids and learn about their jobs in STEM!

Poppy is a nine-year-old girl. She has been to five national parks and hopes to go to visit all 58 of them! Her personal hero is naturalist John Muir (the man who started the National Parks Service and the Sierra Club). Like Muir, she wants to help keep the national parks open for animals to roam free and for people to learn to appreciate nature. “Then, they will fight to protect it!” She hopes to be a park ranger one day.

Malik is a nine-year-old boy. He’s always been interested in lasers, especially the types he sees at concerts and sporting events. He read that lasers are used in tools that can heal hearts and eyes, transmit information around the world, and even cut different materials. He wants to be a photonics engineer and work with lasers one day.

Hannah is a nine-year-old girl. She likes to take things apart, put them back together, and repair things that are broken. She spends a lot of time on her father's job site and watches him cut, shape, and weld things together for all types of industries. She wants to be like her dad and become a welder, too.

Antonio is a ten-year-old boy. He became fascinated with robots ever since he watched Lost in Space reruns with his dad. Recently, he went to a F.I.R.S.T. (For the Inspiration and Recognition of System and Technology) robotics competition where student-built robots competed in games! He wants to be on a FIRST team and become a robotics engineer one day.

Owen is an eight-year-old boy. He loves exploring his backyard for insects. He can't get enough of them! His love of bugs all started when he caught his first firefly. Now, he catches insects with his camera and sketchbook. He has learned how important insects are and how necessary they are to our lives. He wants to become an entomologist when he grows up.

Grace is a ten-year-old girl. She loves computers. She likes to do research, play games, and watch videos on them. She is interested in learning how to write computer code and program computers. She recently read a biography about Grace Hopper, a computer scientist and inventor of COBOL (one of the first high-level programming languages) and wants to be just like her! She hopes to be a computer programmer when she grows up and learn how to program computers to do all kinds of things.

Maya is a ten-year-old girl. She spends most of her time hiking and exploring outdoors with her sister Marisol. On each hike, she collects rocks and pebbles. She has a large collection. She is curious about how rocks form, the layers of the Earth, and fossils, too. She wants to be a geologist one day.

Marisol is a six-year-old girl. She comes from a long line of firefighters and paramedics. She loves spending time at her father's fire station learning how they protect the neighborhood and save lives. She is always ready with a bandage if her sister Maya scrapes her knee on a hike. She wants to follow in her family footsteps and be a paramedic/EMT.

Hiro is an eight-year-old boy and swims like a fish. His favorite things to do are swimming and snorkeling. If he could live in the water, he would! Recently, he was on a trip to Florida and saw all kinds of fish while snorkeling, and went to an oceanarium to learn about the ocean and its animals. He wants to become an ocean engineer to help protect oceans.

Looking at the Back Pages

The Inspire Science Handbook has several resources in the back of the book that will help you learn more about science, technology, engineering, and mathematics. You will find useful information about STEM careers and ways to practice math and reading skills in science, too!

Career Connections

Have you ever wondered what you might do as a career? Careers in STEM are very exciting and are in high demand. These types of careers are important to keep our world safe and clean while also making advancements in technology. Check out the Careers section in the back of the book to learn about some of the awesome STEM careers that are out there in Life, Earth, and Physical science as well as Engineering.

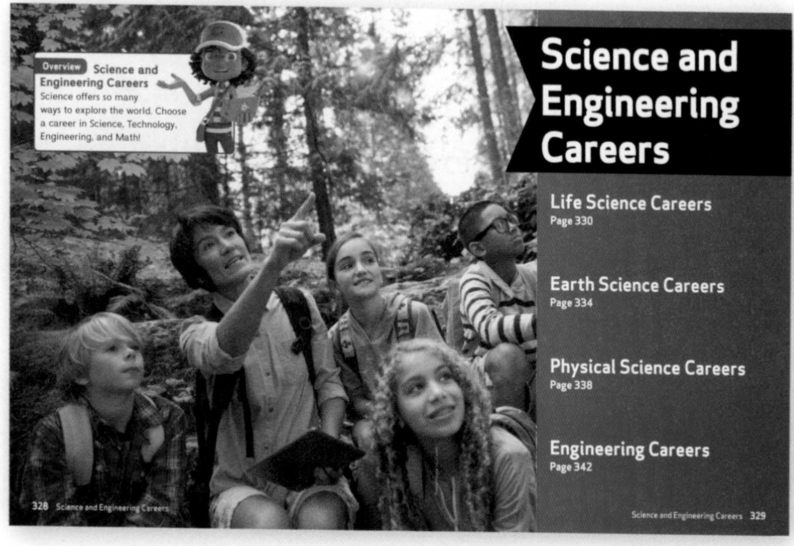

Science Guide

There are many skills that are needed in order to be a scientist or engineer. They even use math and reading skills just like you do in school! The Science Guide pages will give tips on how to use the tools and strategies used by scientists and engineers.

Overview Science and Engineering Guide
Science exploration uses many tools to measure data and to interpet information. Learn about them in this section!

346 Science Guide

Science Guide

Using Maps
Page 348

Using Math
Page 354

Using Language
Page 360

Research Skills
Page 362

Glossary
Page 366

Science Guide 347

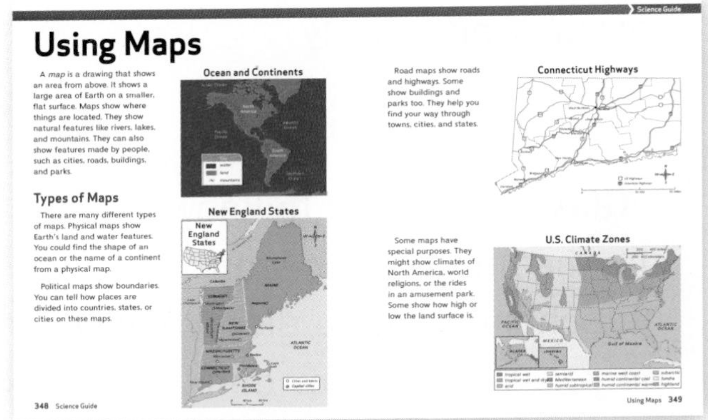

Science Guide

Using Maps

A *map* is a drawing that shows an area from above. It shows a large area of Earth on a smaller, flat surface. Maps show where things are located. They show natural features like rivers, lakes, and mountains. They can also show features made by people, such as cities, roads, buildings, and parks.

Types of Maps

There are many different types of maps. Physical maps show Earth's land and water features. You could find the shape of an ocean or the name of a continent from a physical map.

Political maps show boundaries. You can tell how places are divided into countries, states, or cities on these maps.

Ocean and Continents

New England States

Road maps show roads and highways. Some show buildings and parks too. They help you find your way through towns, cities, and states.

Connecticut Highways

Some maps have special purposes. They might show climates of North America, world religions, or the rides in an amusement park. Some show how high or low the land surface is.

U.S. Climate Zones

348 Science Guide

Using Maps 349

Maps Maps are helpful in knowing your location and the location of other places on Earth. There are several different kinds of maps that can even tell you the climate and landforms found in a region.

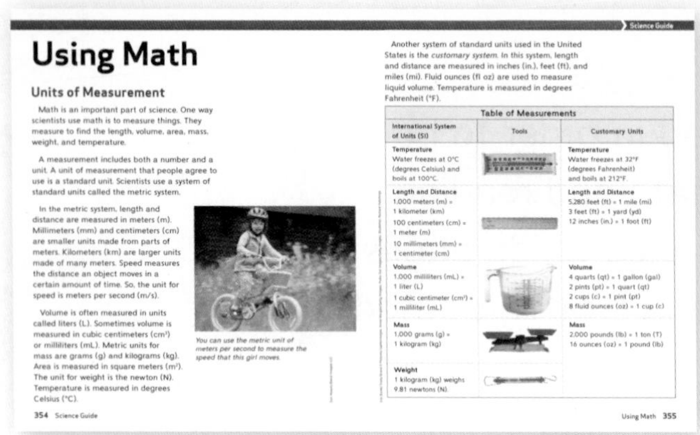

Using Math

Units of Measurement

Math is an important part of science. One way scientists use math is to measure things. They measure to find the length, volume, area, mass, weight, and temperature.

A measurement includes both a number and a unit. A unit of measurement that people agree to use is a standard unit. Scientists use a system of standard units called the metric system.

In the metric system, length and distance are measured in meters (m). Millimeters (mm) and centimeters (cm) are smaller units made from parts of meters. Kilometers (km) are larger units made of many meters. Speed measures the distance an object moves in a certain amount of time. So, the unit for speed is meters per second (m/s).

Volume is often measured in units called liters (L). Sometimes volume is measured in cubic centimeters (cm^3) or milliliters (mL). Metric units for mass are grams (g) and kilograms (kg). Area is measured in square meters (m^2). The unit for weight is the newton (N). Temperature is measured in degrees Celsius (°C).

You can use the metric unit of meters per second to measure the speed that this girl moves.

Another system of standard units used in the United States is the *customary system*. In this system, length and distance are measured in inches (in.), feet (ft), and miles (mi). Fluid ounces (fl oz) are used to measure liquid volume. Temperature is measured in degrees Fahrenheit (°F).

Table of Measurements

International System of Units (SI)	Tools	Customary Units
Temperature Water freezes at 0°C (degrees Celsius) and boils at 100°C.		**Temperature** Water freezes at 32°F (degrees Fahrenheit) and boils at 212°F.
Length and Distance 1,000 meters (m) = 1 kilometer (km) 100 centimeters (cm) = 1 meter (m) 10 millimeters (mm) = 1 centimeter (cm)		**Length and Distance** 5,280 feet (ft) = 1 mile (mi) 3 feet (ft) = 1 yard (yd) 12 inches (in.) = 1 foot (ft)
Volume 1,000 milliliters (mL) = 1 liter (L) 1 cubic centimeter (cm^3) = 1 milliliter (mL)		**Volume** 4 quarts (qt) = 1 gallon (gal) 2 pints (pt) = 1 quart (qt) 2 cups (c) = 1 pint (pt) 8 fluid ounces (oz) = 1 cup (c)
Mass 1,000 grams (g) = 1 kilogram (kg)		**Mass** 2,000 pounds (lb) = 1 ton (T) 16 ounces (oz) = 1 pound (lb)
Weight 1 kilogram (kg) weighs 9.81 newtons (N).		

Math in Science Science and math are very closely related. Math skills are needed for making measurements and analyzing data in experiments and in everyday observations. This section will help you understand the math systems that apply to science and engineering practices.

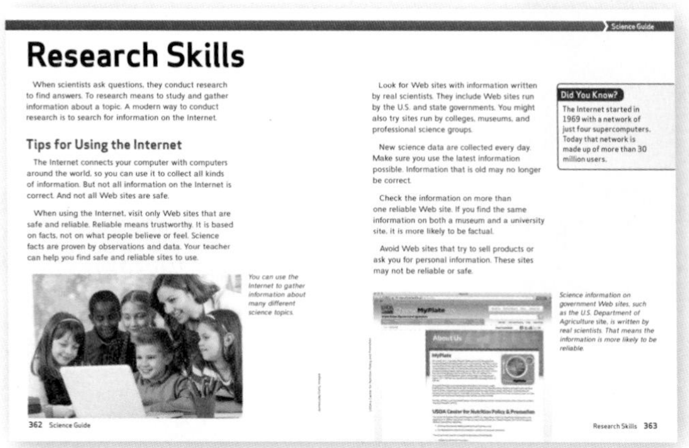

Research Skills

When scientists ask questions, they conduct research to find answers. To research means to study and gather information about a topic. A modern way to conduct research is to search for information on the Internet.

Tips for Using the Internet

The Internet connects your computer with computers around the world, so you can use it to collect all kinds of information. But not all information on the Internet is correct. And not all Web sites are safe.

When using the Internet, visit only Web sites that are safe and reliable. Reliable means trustworthy. It is based on facts, not on what people believe or feel. Science facts are proven by observations and data. Your teacher can help you find safe and reliable sites to use.

You can use the Internet to gather information about many different science topics.

Look for Web sites with information written by real scientists. They include Web sites run by the U.S. and state governments. You might also try sites run by colleges, museums, and professional science groups.

New science data are collected every day. Make sure you use the latest information possible. Information that is old may no longer be correct.

Check the information on more than one reliable Web site. If you find the same information on both a museum and a university site, it is more likely to be factual.

Avoid Web sites that try to sell products or ask you for personal information. These sites may not be reliable or safe.

Did You Know?

The Internet started in 1969 with a network of just four supercomputers. Today that network is made up of more than 30 million users.

Science information on government Web sites, such as the U.S. Department of Agriculture site, is written by real scientists. That means the information is more likely to be reliable.

Research Skills Being able to read, write, and record information is essential to being a scientist or an engineer. Sometimes you will need to research information to help you understand science content and support your observations. This section will provide ways that you can perform research effectively.

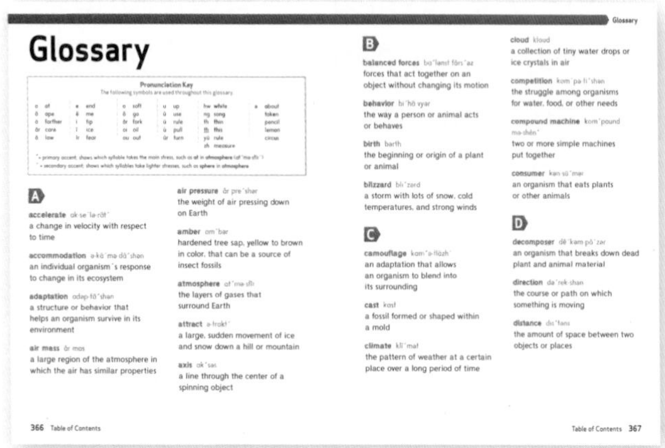

Glossary

Pronunciation Key

The following symbols are used throughout this glossary

A

accelerate ak-se´la-rāt´
a change in velocity with respect to time

accommodation a-kä´ma-dā´shan
an individual organism´s response to change in its ecosystem

adaptation adap-tā´shan
a structure or behavior that helps an organism survive in its environment

air mass âr mas
a large region of the atmosphere in which the air has similar properties

air pressure âr pre´shar
the weight of air pressing down on Earth

amber am´bar
hardened tree sap, yellow to brown in color, that can be a source of insect fossils

atmosphere at´ma-sfir
the layers of gases that surround Earth

attract a-trakt´
a large, sudden movement of ice and snow down a hill or mountain

axis ak´sas
a line through the center of a spinning object

B

balanced forces ba´lanst fôrs´az
forces that act together on an object without changing its motion

behavior bi´hā vyar
the way a person or animal acts or behaves

birth barth
the beginning or origin of a plant or animal

blizzard bli´zard
a storm with lots of snow, cold temperatures, and strong winds

C

camouflage kam´a-fläzh´
an adaptation that allows an organism to blend into its surrounding

cast kast
a fossil formed or shaped within a mold

climate klī´mat
the pattern of weather at a certain place over a long period of time

cloud kloud
a collection of tiny water drops or ice crystals in air

competition kom´pa-ti´shan
the struggle among organisms for water, food, or other needs

compound machine kom´pound ma-shēn´
two or more simple machines put together

consumer kan-sü´mar
an organism that eats plants or other animals

D

decomposer dē´kam-pō´zar
an organism that breaks down dead plant and animal material

direction da´rek-shan
the course or path on which something is moving

distance dis´tans
the amount of space between two objects or places

Glossary Want to know the meaning of the word mitochondria? The glossary will be your resource for definitions of vocabulary words that will be presented in each section. Use the glossary like a dictionary for the highlighted words you encounter as you read the information in the Handbook.

InspireScience

Overview

Science and Engineering Basics

Science follows scientific methods of exploration. These pages will lead you through the steps!

Science and Engineering Basics

What are Science and Engineering?

The fields of science and engineering involve the collection of data and logical thinking. Science and engineering are often used together. The microscope was created using science and engineering. People used science to understand lenses. They used what they learned to produce a tool that used lenses. They used engineering to create the tool. Science and engineering are a part of your daily life. Learning to evaluate data and being able to think logically are important. This section will help you understand methods that scientists and engineers use.

A microscope was created using science and engineering.

Science

The word science may bring to mind microscopes and people in lab coats. These are all parts of science, but what is science, really? Science was being conducted long ago, even before microscopes. The first person who wondered what plants were made of and then tried to figure it out was using science!

Science is trying to answer questions about the world by using evidence.

Scientists learned how lenses and mirrors change the way they saw something. They used what they learned to create the microscope. Microscopes were invented to help people see tiny things. They could not see these things with just their eyes. Using microscopes, they were able to answer more questions about the world. Using science to create tools can make more science possible.

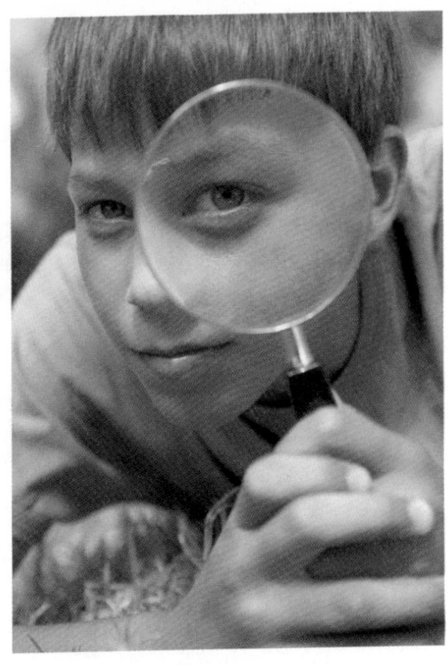

A lens can make something look larger.

However, answering science questions does not create new tools right away. That creation is done through engineering.

Engineering

Engineering is using science and math to solve a problem or need.

Engineers can test ways different lenses work together. Sometimes they make an image larger. Sometimes they make it smaller. An engineer would design and test several different ways the lenses could be set up to find the best way to get the largest image. The best set-up would be used to build a microscope. So science leads to engineering. Answering questions can help solve problems. Solving problems can help answer questions.

Science and Engineering Practices

Scientists and engineers use certain practices when working.

Practices are sets of skills and knowledge.

Many of these practices are things that you do during everyday life, even if you do not realize it. Most of the practices are used in both science and engineering. A few are used just for science or just for engineering. You will learn more about each practice in this section.

A cricket makes a chirping noise by rubbing its wings together.

Asking Questions and Defining Problems

Asking Questions

Asking questions is a science practice that drives every scientific investigation or experiment. Asking questions is not as simple as it may seem. A poorly worded question does not help a scientist gather much information. A good question can lead to important information and connections, or even more questions.

Questions in science can come from observations. *Observation* is using your senses to gather information and take note of what occurs. For example, if you hear a chirping noise, you might ask: "What is that sound?" You can use observations to ask more specific questions. You follow the noise and find that it is coming from a cricket. Then you may ask: "Why is the cricket making that sound? How is it making that sound?" These are clearer questions coming from observation.

Defining Problems

Defining problems is an engineering practice that underlies any solution. Engineers study how people do things and try to make the experience better. If people do not have a way to do something yet, engineers invent it.

Engineers identify problems for people and society and then design solutions to those problems. The solution could be a process, a system, or an object, such as a tool. Space suits worn by astronauts are technological solutions designed by engineers.

Defining a problem involves thinking about all questions, information, and procedures. You need to think about any requirements for the solution. A well-defined problem contains all the requirements for a solution.

A space suit is a solution to the problem of humans being able to survive in space.

NASA/JSC

Developing and Using Models

Defining a Model

You may be familiar with models if you or a friend has ever built a model spaceship, car, or castle with construction toys. What about scientific models? In your science classes, you have probably seen or even made models of the solar system.

A *model* is a representation of a part of an idea, event, process, structure, or object.

Models allow scientists to answer questions and engineers to solve problems because they help people understand how things are made and how they work. Models come in many forms. They can be diagrams, math problems, computer simulations, maps, or physical reproductions.

This model of the solar system shows how the planets orbit around the Sun.

Anthony Bradshaw/Getty Images

Describing the Very Large or the Very Small

Models can describe processes that cannot be observed with our senses. Sometimes processes are too large or too small to see. When you get a cut, you know that a scab forms, but you cannot see the tiny processes going on that help it to form. A digital model can help you visualize what is happening. Many processes in the universe, the solar system, or even on Earth are too big to directly observe. A model helps you see them on a smaller scale.

Generating Data

Models may represent data from an investigation, and they can be used to generate data that are useful for answering questions or finding solutions. *Data* are the facts or information used to figure something out.

Data are often numbers, but they do not have to be. Models can be used to generate data to test ideas. Suppose your class is walking to a field trip location. A computer mapping tool would allow you to put in your destination. Then it would provide you with data: the various choices for routes and the amount of time each will take.

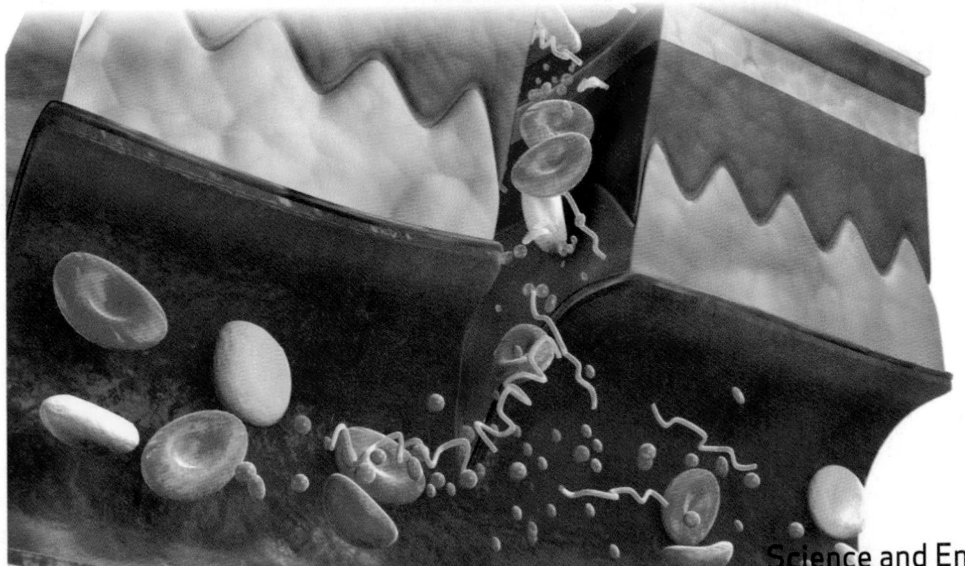

This image of blood clotting comes from a digital model showing what is too small for your eyes to see.

Planning and Carrying Out Investigations

Planning and carrying out investigations is an important science and engineering practice. In fact, you cannot do science without it.

Forming a Hypothesis

Scientists and engineers rely on carefully planned investigations to do their work. First a scientist asks a question or an engineer identifies a problem. Next he or she forms a hypothesis based on experience and research. A *hypothesis* is an explanation of something that can be tested.

An investigation must be done to test the hypothesis. During the investigation, the scientist or engineer will collect data. The data can be used to determine if the hypothesis is proven or if it needs to be changed. The investigation process can start all over based on the new information. Scientists and engineers often use the steps shown to the right when carrying out investigations.

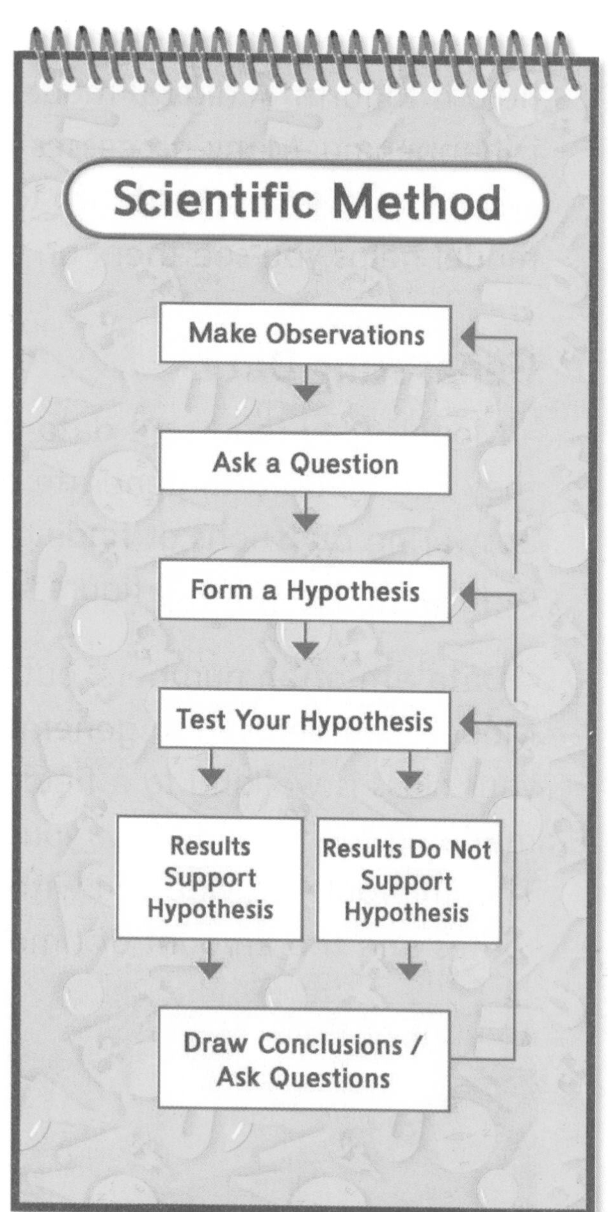

Components of an Investigation

When planning an investigation, the first step is identifying the variables. Variables are the things that may change in the experiment. A good experiment includes three types of variables. Independent variables are the things that you change to test your hypothesis. Dependent variables are the things that change as a result of the changes to the independent variables. Controlled variables are the things you keep the same in each test. The dependent variable is measured or observed. The independent variable is tested or changed to see how it affects the dependent variable.

For example, you could do an experiment to determine what amount of water helps a plant grow the best. You want to make sure the only different between the plants is the amount of water they receive. You control all the other variables. The plants should get the same amount of sunlight, the same temperature, and the same type of soil. These are controlled variables. The independent variable is the amount of water, each plant should receive a different measured amount. The dependent variable is how much each plant grows in a certain amount of time.

These plants grew differently because the amount of water they received was different.

Analyzing and Interpreting Data

Analyzing and interpreting data is very important to scientists and engineers. The data will help answer a question or determine if a solution solves a problem. Scientists and engineers use data to understand relationships between events. Being able to understand these relationships requires knowledge of how to interpret the data correctly.

Using Graphs

Graphs are often useful for analyzing, interpreting, and communicating data. They give a visual representation of how the data is related. Graphs work best when the data collected uses numbers or measurements. You can learn more about using graphs in the Doing Science and Engineering section.

How Variables Relate

One thing that the data will show is the relationship between the variables in the experiment. The dependent variables will change because the independent variable has been changed. The data that you gather should show this relationship. You can compare data from investigations with other variables to look for patterns.

You can use graph paper to make a graph.

Evaluating Data

If you do not collect data accurately, you cannot be sure they are reliable. Measurements need to be precise. When you fill a beaker or a flask, you need to read the level of the liquid correctly. If an investigation involves timing an event to the nearest second, you would likely use a stopwatch. Careful observation is needed so that you stop the timer at exactly the right moment. An electronic timer that uses a laser as a sensor could provide better accuracy, but it would likely be more expensive.

How else could you make your time measurements more accurate? A greater number of trials can improve accuracy. Collecting a large data sample allows you to ignore any numbers that are very different from the majority because you can assume they are inaccurate.

Make sure to measure liquids correctly.

A stopwatch can measure time accurately.

Using Math and Computational Thinking

Math is used in science and engineering to represent variables and their relationships. It is also used to make predictions about events and results. Engineers apply math equations and results to design systems. Scientists use computers and programs developed by engineers. Both fields work together to advance technology and achieve results. Using math and computational thinking is a key practice for both science and engineering.

Using Math

Math is the study of amount, structure, and change. It uses numbers to describe amounts and relationships of objects, processes, and events. The ideas and equations used in math are developed to explain the world around us and everything that happens. It could be said that mathematics is the "language" of science.

Everything in the world around you can be represented using math. Shopping malls, football stadiums, and shoe boxes can be represented using math by their measurements of length and height, which can then be used to calculate their areas and volumes.

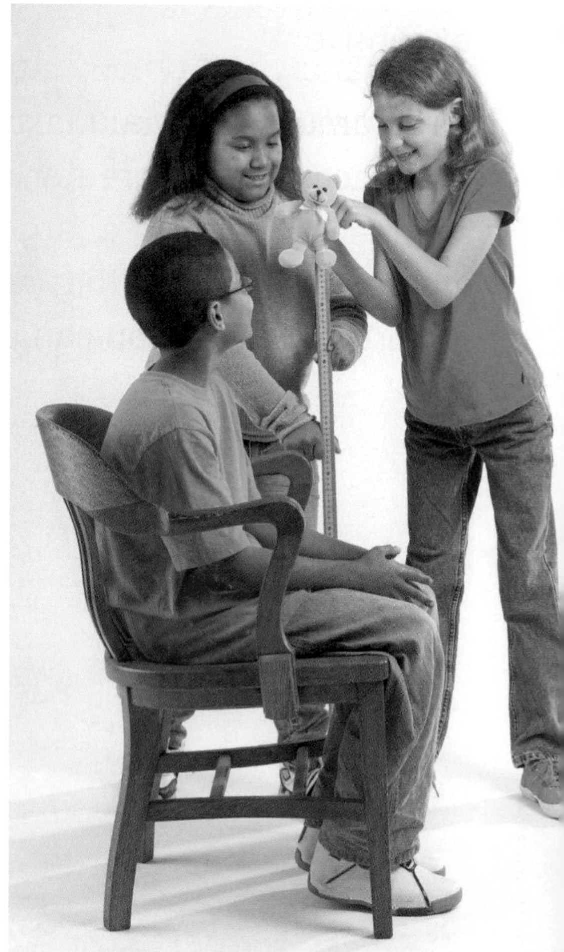

Measuring something allows you to think about it using math. These students are measuring how far the stuffed animal will fall when they drop it.

Richard Hutchings

Computational Thinking

The term *computational* refers to problem-solving using computer-science reasoning. Scientific investigations and engineering designs often rely on computer calculations and simulations. These computer programs need to be developed, tested, and implemented, all of which require computational thinking.

Computational thinking involves the ability to recognize the relationship between events and mathematics. This sort of thinking can allow you to collect useful data and do an analysis. Computer programs can make analyzing large amounts of data easier.

This computer program simulates global winds. It allows scientists to make predictions about the affects those winds will have.

William Putman/NASA Goddard Space Flight Center

Constructing Explanations and Designing Solutions

Constructing Explanations

Once a scientist has asked a question, conducted an investigation, and analyzed the data, his or her goal is to come up with an explanation. The explanation must be based on evidence produced from the investigation. Constructing explanations is a science practice that relates directly to the practice of asking questions.

To explain how something works, scientists must identify how variables relate to each other. Understanding these relationships and considering other factors that could affect the results enable scientists to make valid explanations.

Your explanation can also be supported by existing scientific laws and theories. You can use the data you have collected from experiments to show evidence of these existing science ideas.

You can use the law of gravity to help explain why different objects fall at different speeds.

Jill Braaten/McGraw-Hill Education

Designing Solutions

Recall that after conducting research and analyzing the data, an engineer's ultimate goal is to design a solution to a problem. The solution must be based on evidence from research. Its reliability is based on testing and revising solutions while following design specifications and constraints, or limits. Designing solutions is an engineering practice that relates directly to the practice of defining problems.

Designing a solution usually involves meeting certain requirements. There may also be limits on the project, such as a certain cost or time. A successful solution will meet those requirements within the limits of the project.

An electric car is the result of a design solution. It is a car that does not use as much gas, but still can be made in a cost effective manner.

Engaging in Argument from Evidence

Have you ever gotten into an argument with a sibling or a friend? In everyday life, arguments are often negative events because they are linked with angry or sad emotions. In science and engineering, though, arguments are necessary for advancements to be made. In science, an *argument* is a statement based on logic or evidence that supports or opposes something.

Argumentation ensures that all aspects of an explanation or design solution are considered. Engaging in argument from evidence is a science and engineering practice that serves to identify the best explanation or solution.

You should support your argument with evidence from your experiments or trials.

The Right Way to Argue

People often have opposing views on how best to conduct an investigation, interpret results, or communicate information. These views should have reliable data and scientific principles as evidence in order to be valid. When you plan and conduct an investigation, it is important to evaluate procedures and results, keeping in mind that you need to defend, or give logical reasons for, your choices and results.

The practice of science includes arguing for constructed explanations, defending interpretations of data, and recommending proposed designs. Real-world solutions to problems have to consider constraints and tradeoffs that impact society. Often, solutions lead to other problems that are not easy to predict. It is important that you listen to, compare, and evaluate competing ideas and methods based on their values and limitations.

Working with others can help you improve your argument.

Kali Nine LLC/Getty Images

Obtaining, Evaluating, and Communicating Information

Communication is important in every aspect of our lives. In science and engineering, it allows for new explanations and design solutions to be shared with society and the science and engineering communities. These communications can be written or verbal. They can be online, in books, on the news, or in public or private discussion. Learning how to understand and evaluate communicated information is valuable if you are interested in pursuing a career in science.

Obtaining Information

Scientific information is valuable not only for scientists and engineers, but also for the general public. Integrating information from various media sources helps to give you a better understanding. Types of *media* include written texts, such as newspapers and Internet articles, as well as radio, television, and videos.

You can use internet or book resources to gather information.

Evaluating Information

You must think critically about the scientific information you read in order to evaluate it. When obtaining scientific information, you should critique the source to determine if it is credible or biased. Information is considered *credible* if it can be trusted to be true. *Biased* information reflects a strong interest of the writer or sponsor. You need to evaluate the information you gather and decide whether more information is needed.

Share information that you have obtained with others.

Communicating Information

Communicating the information you have obtained and evaluated will help to educate others. This communication will include sharing the findings of your own investigations as well.

Science and Engineering Processes

The Scientific Process

Scientists all over the world follow a similar process when they are investigating or trying to find a solution to a problem. This process is known as the scientific process and allows scientists and engineers to maintain consistency and reliability in their findings.

The Steps of the Scientific Process

Make Observations One of the most important skills a scientist must have is to be able to make careful observations of the world around them. This step of the scientific process is a starting point for scientists to find reasons to explore new ideas.

Ask a Question The scientific process as a whole is driven by a question that the scientist or engineer is trying to find the answer to. This question comes from a scientist's observations of something they want to know more about or from a problem that needs to be solved. For example, a scientist might ask, "How can we use biomass as fuel for cars?," and an engineer might ask, "What material is best to build a skyscraper near the ocean?"

Form a Hypothesis Scientists and engineers need to form a hypothesis when investigating a problem or new idea. A hypothesis is a statement that can be tested to answer a question. An educated prediction can guide the scientist to support or disprove it based on their findings.

Test the Hypothesis This part of the scientific process has several steps. First, scientists have to select a strategy for how they will test their hypothesis. They might plan an experiment or collect information through a survey, depending on the topic. Then, scientists need to plan their procedure. This procedure needs to be clear enough for other scientists to follow in case the experiment needs to be recreated. Lastly, scientists carry out their plan to test their hypothesis and record data and observations throughout their plan.

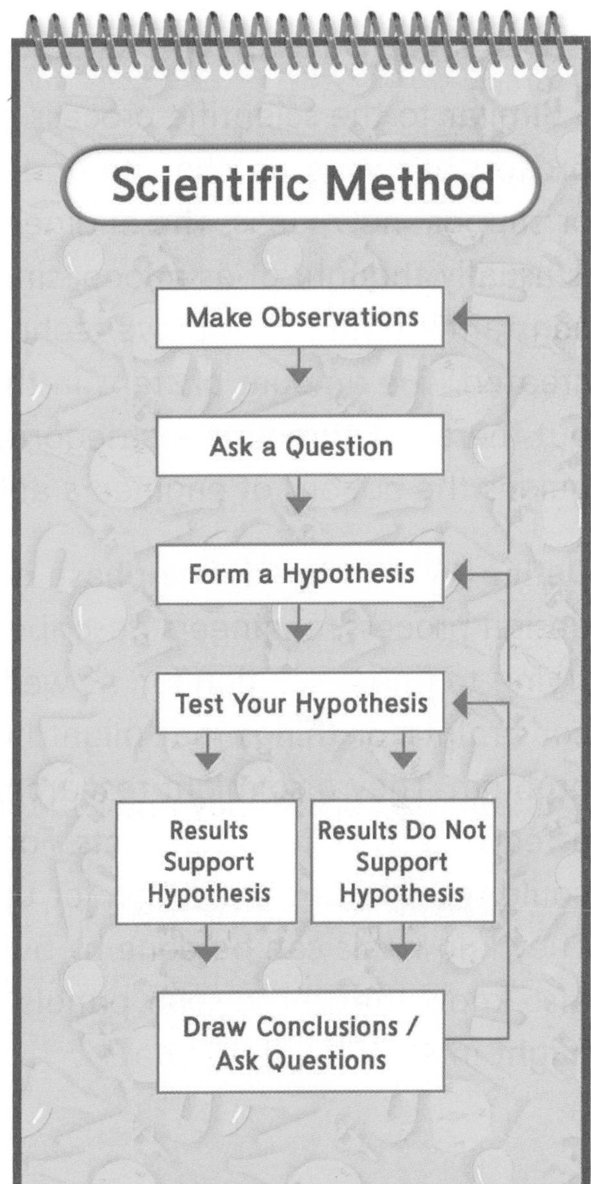

Using Results to Draw Conclusions After scientists finish collecting data by testing their hypothesis, they need to analyze the data to decide if their hypothesis was or was not supported by the results. If their hypothesis is supported by the results, they can move on to draw conclusions about how it might answer their question. If their hypothesis is not supported by the results, they need to go back and come up with a new question or hypothesis to try and get the information they need.

The Engineering Design Process

Similar to the scientific process, the engineering design process allows researchers to find solutions to problems or support new ideas. The engineering design process is usually thought of as a loop, since engineers are constantly trying to improve technologies once they are created. The amount of steps in this process can vary, but there are three main categories in the process that guides the actions of engineers and other scientists.

Define Within the define phase of the engineering design process, engineers describe the problem they are trying to find a solution for, as well as defining possible constraints, or things that might limit a solution to the problem. They also might research how the problem affects many different aspects. For example, engineers could want to find a solution for using less fossil fuel. They know this can be done by building solar panels, but also know that those solar panels are very expensive and might interfere with habitats.

Develop Solutions The next phase of the engineering design process involves researching and testing multiple solutions that might be able to solve the problem. By exploring multiple solutions, engineers can weigh their options in terms of materials, cost, and other factors. Engineers usually work as a team to build models and test their designs, recording data as they go so that they can make informed decisions.

Optimize Solutions The engineering design process is one that is continuous and focuses on improving solutions to make them better with each iteration. This phase involves improving successful solutions based on tests. Even if a solution fails, engineers and other scientists can use this information to improve the design! Often times, even simple solutions need to be revised several times before it is reliable enough to use again and again.

The Engineering Design Process

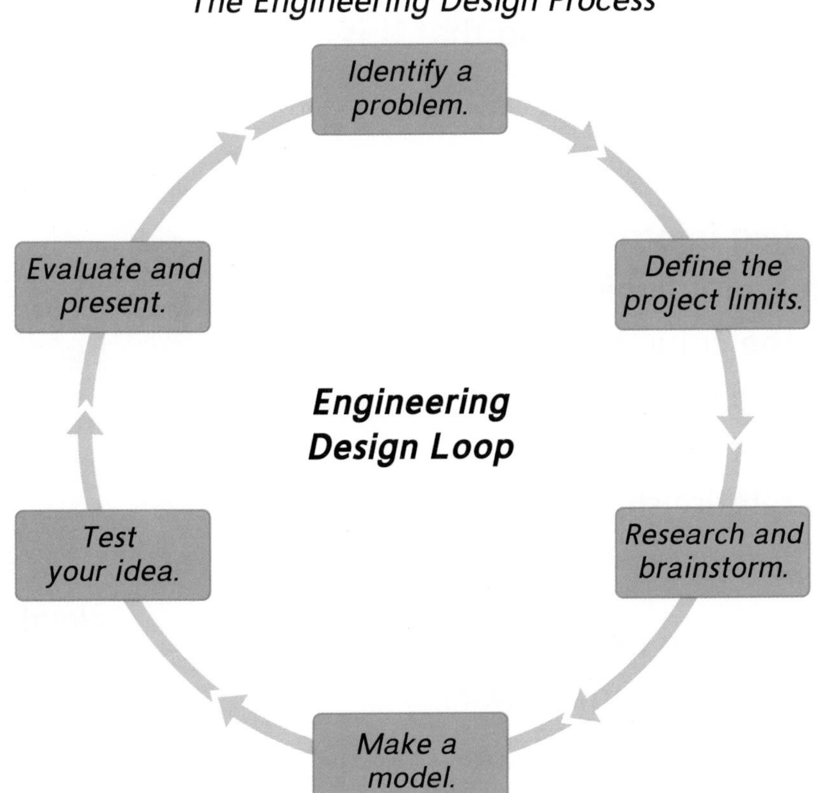

Doing Science and Engineering

In order to put science and engineering into practice, it is important to have several skills that integrate organizing data, working safely, and using tools that will result in the information needed to solve a problem or positively influence society.

Organizing Data

It is important to organize any data that you find through science and engineering investigations so that it can be communicated to others. Organizing data can also help you find patterns or trends that may tell you something new about what you were trying to find. There are many ways that data can be organized so that it is the most useful.

Make a Chart

Charts are useful for recording information during an experiment and for communicating information. In a chart, only the column or row has meaning, but not both. In this chart, one column lists living things, while the second lists nonliving things. There is no relationship across the rows of objects.

Living	Nonliving
tree	rock
chipmunk	puddle
bird	cloud

Organizing data from a survey can be done with a chart. The question and corresponding answers can be displayed in the columns and rows.

Make a Table or Web

Tables Tables can help you organize and record data during an investigation. The columns and rows have headings that tell you what kind of information goes in each part. The table below shows the properties of certain minerals. Follow each row and column to learn about each mineral's characteristics.

Mineral Identification Table					
	Hardness	Luster	Streak	Color	Other
pyrite	6–6.5	metallic	greenish-black	brassy yellow	called "fool's gold"
quartz	7	nonmetallic	none	colorless, white, rose, smoky, purple, brown	
mica	2–2.5	nonmetallic	none	dark brown, black, or silver-white	flakes when peeled
feldspar	6	nonmetallic	none	colorless, beige, pink	
calcite	3	nonmetallic	white	colorless, white	bubbles when acid is placed on it

Idea Webs An idea web shows how ideas or concepts are connected to each other. Idea webs help you organize information about a topic. This idea web shown here connects different ideas about rocks.

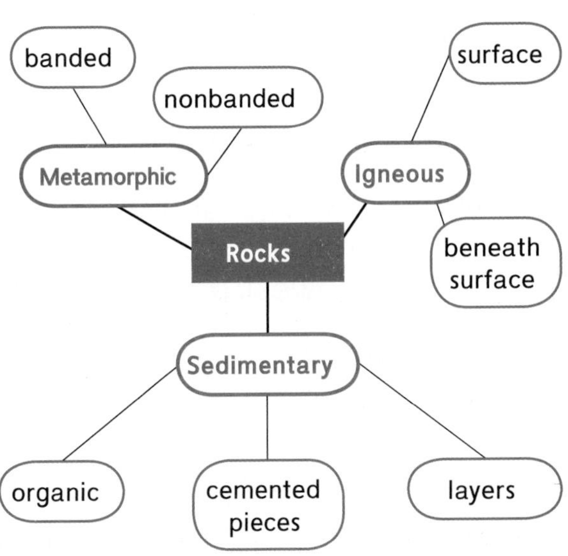

Organizing Data

Make a Graph

Graphs are very helpful in analyzing data that has been organized. This next level of organization can make it easy to notice patterns, trends, and relationships within the data that was collected.

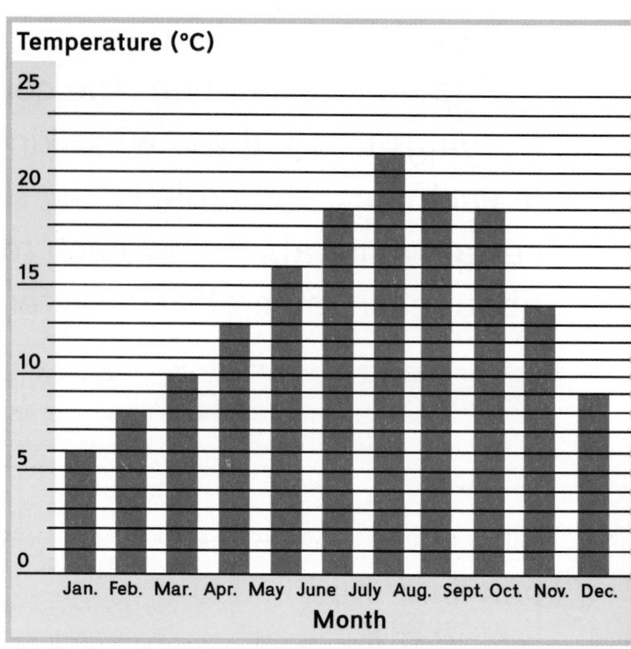

Bar Graphs A bar graph uses bars to show the quantity of a type of data. The warmest and coldest months for a city can be graphed. Every month, the average temperature can be found online or in the newspaper. The data can then be organized in a chart and then used to make a bar graph.

Month	Temperature (°C)
January	6
February	8
March	10
April	13
May	16
June	19
July	22
August	20
September	19
October	14
November	9
December	7

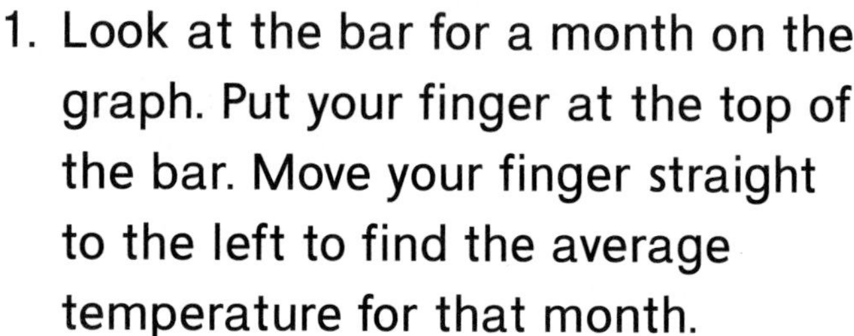

1. Look at the bar for a month on the graph. Put your finger at the top of the bar. Move your finger straight to the left to find the average temperature for that month.
2. Find the highest bar on the graph. This represents the month with the highest average temperature.
3. Find the lowest bar on the graph. This represents the month with the lowest average temperature.
4. Do you see any patterns in the average monthly temperature for this town?

Pictographs A pictograph uses symbols or pictures to show quantities of information. It is different from a bar graph because it shows data in increments. For example, estimated water usage each day can be shown like in the pictograph below.

Water Used Daily (liters)	
drinking	10
showering	100
bathing	120
brushing teeth	40
washing dishes	80
washing hands	30
washing clothes	160
flushing toilet	50

Line Graphs A line graph can show how information changes over time. Information can be recorded over the course of an hour or day or even over a year or decade. Temperature is often recorded over time for weather forecasts and can then easily be represented as a line graph.

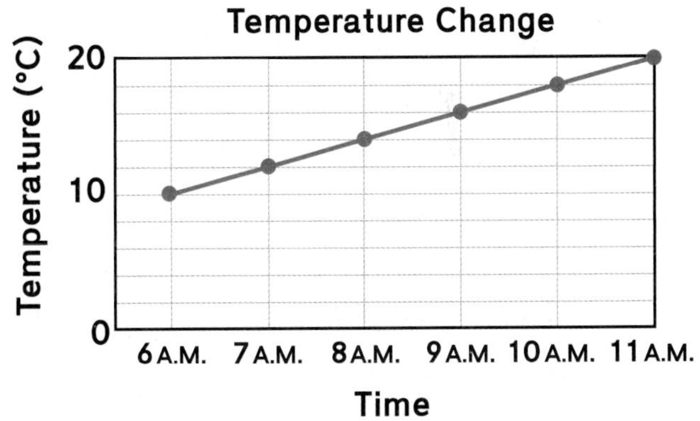

Time	Temperature (°C)
6 A.M.	10
7 A.M.	12
8 A.M.	14
9 A.M.	16
10 A.M.	18
11A.M.	20

Working Safely

Think about the times in your life when there were rules that you needed to follow. We know that rules are important to keep us safe and to keep situations fair. It is also very important for scientists and engineers follow certain safety rules while they do their jobs. Safety rules are necessary to make sure that no one becomes ill or injured. Safety rules are important whether they are applied in a classroom. laboratory setting, or in the field.

In the Classroom

Before an Activity

- Always read all of the directions before starting an investigation. Make sure you understand them. If you have questions about the directions, ask your teacher to help you understand.
- When you see a warning in the directions, such as BE CAREFUL, be extra careful to follow all of the safety rules.
- Listen to your teacher for special safety directions. If you do not understand something, ask for help.
- Wash your hands with soap and water before (and after!) an activity.
- Always pay attention to your surroundings when you gather materials for the activity.

During an Activity

- Wear a safety apron if you work with anything messy or with something that might spill.
- Wear safety goggles when your teacher tells you to wear them. Wear them when working with anything that can fly into your eyes or when working with liquids that might splash.
- Keep your hair and clothes away from materials. Tie your hair back if it is long and roll up long sleeves that might get in the way.
- Do not eat or drink anything during the experiment. A material may be safe to touch, but it may be dangerous to eat.

After an Activity

- Dispose of materials the way your teacher tells you to.
- Put equipment and extra materials back following your teacher's instructions.
- Clean up your work area and wash your hands.

Working Safely

Special Safety Rules

Sometimes, science and engineering investigations require special equipment or for you to study something outside the classroom. Keep these safety rules in mind along with the ones that you saw on the previous pages.

Handling Materials

- If something spills, clean it up right away so that no one slips and falls and the material does not damage anything. If you are not sure how to clean something up correctly, ask your teacher.
- Tell your teacher if something breaks. If glass breaks, do not clean it up yourself.
- Keep your hands dry when using equipment that uses electricity.

Outside the Classroom

- Always travel outside the classroom with a trusted adult such as your teacher, parent, or guardian.
- Do not touch animals or plants without an adult's approval. The animal might bite, and the plant might be poisonous.
- Remember to treat living things, the environment, and one another with respect.

(t) Ken Karp/McGraw-Hill Education; (b) Dave & Les Jacobs/Getty Images

Measurements

Units of Measurement

Temperature The temperature on this thermometer reads 46 degrees Fahrenheit. That is the same as 8 degrees Celsius.	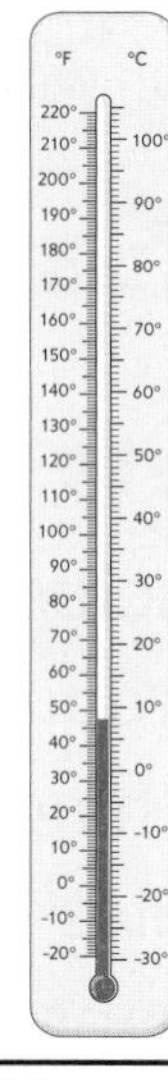
Length This girl is 4 feet and 11 inches. That is the same 1 meter and 50 centimeters	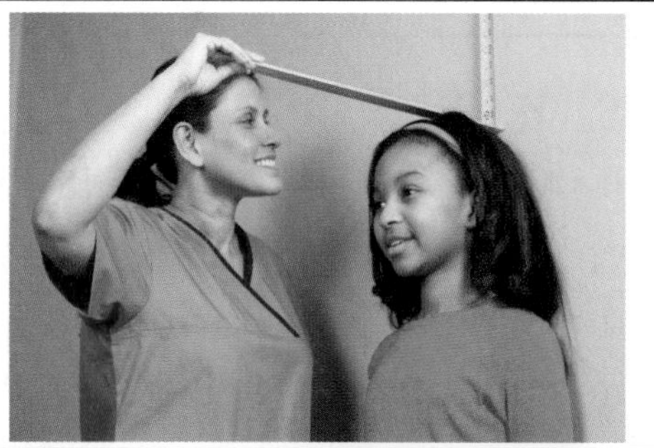
Mass You can measure the mass of this box of crayons in grams.	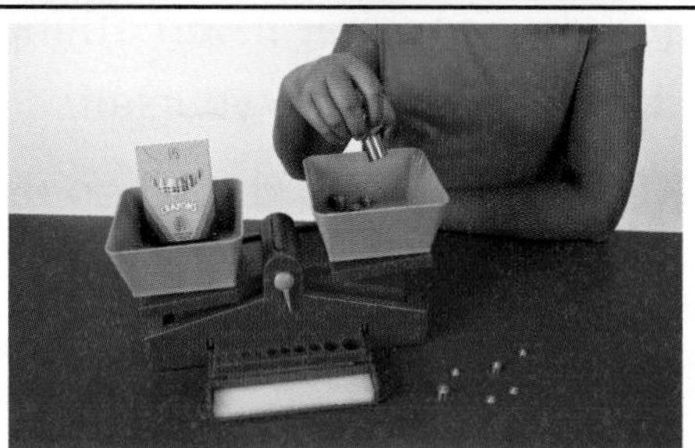
Volume of Fluids This bottle of liquid has a volume of 2 liters. That is a little more than 2 quarts. This carton holds 1 quart of milk.	

Weight/Force You would need 11.8 newtons to move this little bowl of fruit which weighs only 2.5 pounds.	
Speed This woman can ride her bike 100 meters in 50 seconds. That means her speed is 2 meters per second.	

Table of Measures

SI International Units/Metric Units	Customary Units
Temperature Water freezes at 0 degrees Celsius (°C) and boils at 100°C.	**Temperature** Water freezes at 32 degrees Fahrenheit (°F) and boils at 212°F.
Length and Distance 10 millimeters (mm) = 1 centimeter (cm) 100 centimeters = 1 meter (m) 1,000 meters = 1 kilometer (km)	**Length and Distance** 12 inches (in.) = 1 foot (ft) 3 feet = 1 yard (yd) 5,280 feet = 1 mile (mi)
Volume 1 cubic centimeter (cm^3) = 1 milliliter (mL) 1,000 milliliters = 1 liter (L)	**Volume of Fluids** 8 fluid ounces (fl oz) = 1 cup (c) 2 cups = 1 pint (pt) 2 pints = 1 quart (qt) 4 quarts = 1 gallon (gal)
Mass 1,000 milligrams (mg) = 1 gram (g) 1,000 grams = 1 kilogram (kg)	**Area** 1 square foot (ft^2) = 1 ft x 1 ft 43,560 square feet (ft^2) = 1 acre
Area 1 square meter (m^2) = 1 m x 1 m 10,000 square meters (m^2) = 1 hectare	**Speed** miles per hour (mph)
Speed meters per second (m/s) kilometers per hour (km/h)	**Weight/Force** 16 ounces (oz) = 1 pound (lb) 2,000 pounds = 1 ton (T)
Weight/Force 1 newton (N) = 1 kg x $1 m/s^2$	

Measurements

Measure Time

You measure time to find out how long something takes to happen. Stopwatches and clocks are tools you can use to measure time. Seconds, minutes, hours, days, and years are some units of time.

Push the button on the top right of the stopwatch to start timing. Push the button again to stop timing.

Use a Stopwatch to Measure Time

Get a cup of water and an antacid tablet from your teacher. Tell your partner to place the tablet in the cup of water. Start the stopwatch when the tablet touches the water. Stop the stopwatch when the tablet completely dissolves. Record the time shown.

Measure Length

You measure length to find out how long or how far away something is. Rulers, tape measures, and meter sticks are some tools you can use to measure length. You can measure length using units called meters. Smaller units are made from parts of meters. Larger units are made of many meters.

Look at a ruler. Each number represents 1 centimeter (cm). There are 100 centimeters in 1 meter. In between each number are 10 lines. The distance between each line is equal to 1 millimeter (mm). There are 10 millimeters in 1 centimeter.

Measure Liquid Volume

Volume is the amount of space something takes up. Beakers, measuring cups, and graduated cylinders are tools you can use to measure liquid volume. These containers are marker in units called milliliters (mL).

Measurements

Measure Mass

Mass is the amount of matter an object has. You use a balance to measure mass. To find the mass of an object you compare it with objects whose masses you know. Grams are units people use to measure mass.

Measure Force/Weight

You measure force to find the strength of a push or pull. Force can be measured in units called newtons (N). A spring scale is a tool used to measure force.

Weight is a measure of the force of gravity pulling down on an object. A spring scale measures the pull of gravity. One pound is equal to about 4.5 N.

Measure Temperature

Temperature is how hot or cold something is. You use a tool called a thermometer to measure temperature. In the United States, temperature is often measured in degrees Fahrenheit (°F). However, you can also measure temperature in degrees Celsius (°C).

McGraw-Hill Education/Ken Karp

Tools of Science

Use a Microscope

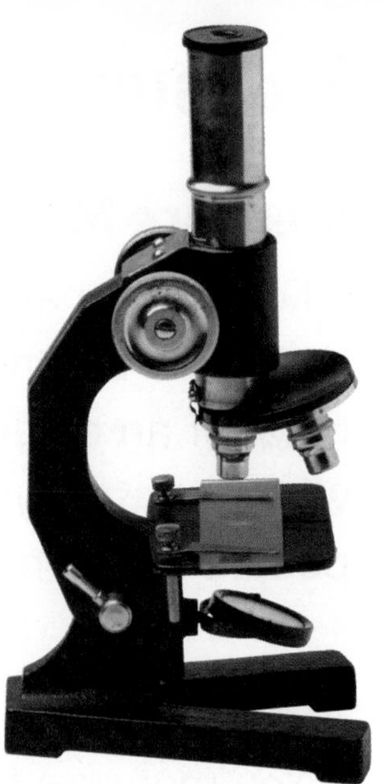

A microscope is a tool that magnifies objects, or makes them look larger. A microscope can make an object look hundreds or thousands of times larger. Look at the photo to learn the different parts of a microscope.

Use a Hand Lens

A hand lens is another tool that magnifies objects. It is not as powerful as a microscope. However, a hand lens still allows you to see details of an object that you cannot see with your eyes alone. As you move a hand lens away from an object, you can see more details. If you move a hand lens too far away, the object will look blurry.

Use a Calculator

Sometimes during an experiment, you have to add, subtract, multiply, or divide numbers. A calculator can help you carry out these operations.

(t) Photodisc/Getty Images; (c) Joe Polillio/McGraw-Hill Education; (b) McGraw-Hill Education

Tools of Science

Use a Camera

During an experiment or nature study, it helps to observe and record changes that happen over time. Sometimes it can be difficult to see these changes if they happen quickly or very slowly. A camera can help you keep track of visible changes. Studying photos can help you understand what happens over the course of time.

Use a Computer

A computer has many uses. You can use a computer to get information from compact discs (CDs), digital video discs (DVDs), and jump drives. You can also use a computer to write reports and to show information.

The Internet connects our computer with computers around the world, so you can collect all kinds of information. When using the Internet, visit only Web sites that are safe and reliable. Your teacher can help you find safe and reliable sites to use. Whenever you are online, never give any information about yourself to others.

Crosscutting Concepts

There are certain themes that appear throughout science and engineering called crosscutting concepts. They are themes that "cut across" different disciplines, tying them together in various ways. There are seven major crosscutting concepts of science and engineering.

Patterns You see a pattern when something appears or occurs over and over again in particular conditions. This can allow you to predict what will happen under similar circumstances. The seasons change in a repeating pattern according to where Earth is in its orbit. Knowing this helps scientists predict whether other planets have similar seasons based on their tilt and orbit.

Cause and effect The concept of cause and effect, or A leading to B, is an important one in science and engineering. A cause-and-effect relationship may be as simple as a bowling ball knocking over pins or as complex as evaluating how a volcanic eruption can affect weather, soil, organisms, and the atmosphere.

Scale, proportion, and quantity Because much of science studies what occurs at the cellular, molecular, or atomic level, knowledge of scale and proportion is important. DNA is so small that a scale model of it is helpful in understanding its structure. Scale is also important when studying large systems like the solar system. Engineers build models to scale in order to get realistic results from their tests.

Systems and system models Systems show how things interact with each other. The body's organ systems, ecosystems, the Sun-Earth-Moon system, and complex machines such as bicycles are examples of systems found throughout science. System models are used in both engineering and science. People can manipulate different components of a system to see how the rest of the components are affected. For instance, an engineer might make a model of a transportation system to examine traffic flow at different times of the day.

Energy and matter: Flows, cycles, and conservation Energy and matter are interwoven into nearly every aspect of science. Energy flows and matter cycles, and they are both conserved as they do so. Scientists observe how the Sun's energy flows through food webs and how water cycles between Earth and the atmosphere.

Structure and function Learning how something works involves understanding what structures comprise it and how each structure functions. Structure and function can apply to the very large (galaxies) and the very small (atoms). An engineer developing a prosthetic limb must understand the function of each structure in the natural limb in order to mimic its flexibility and strength.

Stability and change One of the aspects of investigating systems is whether they are stable or changing. A system can be said to have stability if most components of it are unchanging—and possibly even if they are changing, but in a predictable, cyclic way. The system changes when something upsets the stability. A system's stability might be importannt to engineers working at the citywide or countrywide level, as a change in the system could affect millions of people.

Future Career

Park Ranger

Do you like to interact with nature? Park rangers get to call a state or national park their office! Park rangers are outside most of the day. They check the health of ecosystems in the park. They also make sure habitats are clean and safe. They may need to report damage in the area and watch carefully for change. Park rangers guide field trips and create events to educate students and other people about the plants and animals in the park. They may also help plant new trees and watch for hunting and fishing in the area.

Life Science

Basics of Life

Characteristics of Living Things

Living things are called **organisms.** Plants and animals are organisms. So are bacteria and other tiny creatures. All living things are alike in some important ways.

All Living Things Grow

An apple seed sprouts and begins to grow into a tree. Over time the tree grows taller. Its trunk grows harder and thicker. Eventually, the tree produces new apples. All organisms use energy to grow. To grow means to change with age.

All Living Things Respond

All living things have ways to respond, or react, to the world around them. Responding helps an organism meet its needs. A plant on a windowsill will grow in the direction of sunlight. The plant needs sunlight to make food. A mouse responds to a prowling cat by dashing into a safe hole.

Trees respond to cooler weather and less daylight during fall. Their leaves change color and fall off.

All Living Things Reproduce

Living things reproduce. To **reproduce** means they make more of their own kind. An oak tree produces acorns that will grow into new trees if conditions are right. An alligator makes a nest and lays eggs in it. Young alligators hatch from those eggs. A wolf gives birth to cubs that grow to be wolves like their parents.

Fact Checker

Some nonliving things seem like they are alive. Fire moves. It seems to grow. But fire does not come from another living thing, so fire is not alive.

Nonliving Things

Nonliving things are different from living things. Some examples of the nonliving things in nature are water, rocks, and soil. People make many nonliving things, such as plastic bottles, metal cans, and glass windows. None of these things can grow, respond, or reproduce. They are not alive.

Skinks reproduce by laying eggs. New skinks will hatch from these eggs.

(l) smileitsmcheeze/iStock/Getty Images; (r) ZoonarAlkimson/age fotostock

Needs of Living Things

All organisms need certain things to stay alive. A lion, for example, needs food, water, air, and space. A tomato plant has the same kinds of needs. A living thing cannot survive if its needs are not met.

Living things get everything they need from their environment. An **environment** is all the living and nonliving things that surround an organism.

These plants take in water through their roots from the soil.

Living Things Need Food

Food gives organisms the energy to grow, respond, and reproduce. Plants take in energy from the Sun. They use that energy to make their own food. Animals eat other organisms for food. Mushrooms are neither plants nor animals. Mushrooms are a type of fungus. They get energy from bits of dead organisms in the soil.

Living Things Need Water

The bodies of organisms are full of water. Even in the desert, the stem of cactus has water inside. Water helps living things break down food, move food inside their bodies, and get rid of waste.

Did You Know?

A few plants, such as the Venus flytrap, capture insects! They use the materials that make up the insects' bodies as well as energy from the Sun to help them grow.

Living Things Need Gases

Animals need a gas called *oxygen*. You cannot see oxygen, but no animal can live long without it. Dogs and birds breathe in oxygen from the air. So do some water animals, such as whales. Fish and most other water animals get oxygen from the water around them.

Plants need oxygen, but they also need a gas called *carbon dioxide*. This gas is also found in air and water. Using energy from sunlight, plants change carbon dioxide and water into food.

Living Things Need Space

All organisms need enough space, or room, to grow and to find food and water. Some living things need a lot of space. An eagle soars far and wide searching for food. On the other hand, goldfish can live in small ponds. Even plants need enough room for their roots to spread out and their leaves to catch sunlight.

This manatee must lift its nose above the surface of the water to breathe in oxygen.

Jim Reid/USFWS

Living Things Are Made of Cells

All living things are made of one or more cells. *Cells* are the building blocks of life. Most cells are so small that you need a microscope to see them. Even a tiny ant has millions of cells in its body.

Although cells are very small, there are some organisms that are made of only one cell. Bacteria, for example, are one-celled organisms. There are many different kinds of bacteria, and they live almost everywhere. Bacteria live in soil and water, in food, and in our bodies.

The chloroplasts in a plant cell, like the cells in this lily, take in energy from sunlight.

Fact Checker

Some kinds of bacteria make us sick, but others are helpful. For example, we use bacteria to make yogurt and cheese. Bacteria also live inside you! They help your body break down the food you eat.

What Could I Be?

Bacteriologist

Interested in becoming an expert on single-celled living things? A bacteriologist studies effects of these tiny organisms on plants and animals. Learn more in the Careers section.

Different Kinds of Cells

If everything in your body is made of cells, why are some body parts different from others? The answer is that the cells are different. When an organism has many cells, the cells can do different jobs.

Think about a lily. Its roots grow underground. The job of the roots is to take in water. Root cells do not make food, so they don't have parts for that job. Leaves do make food for the plant. They have the green parts that take in energy from sunlight.

Animal cells are different, too. The cells in your brain have parts and shapes that help them send and receive messages. Your blood cells have parts and shapes that help them move gases around your body and fight disease. Muscle cells have parts and shapes that allow the body to move. Each type of cell does a different job.

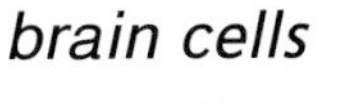

brain cells

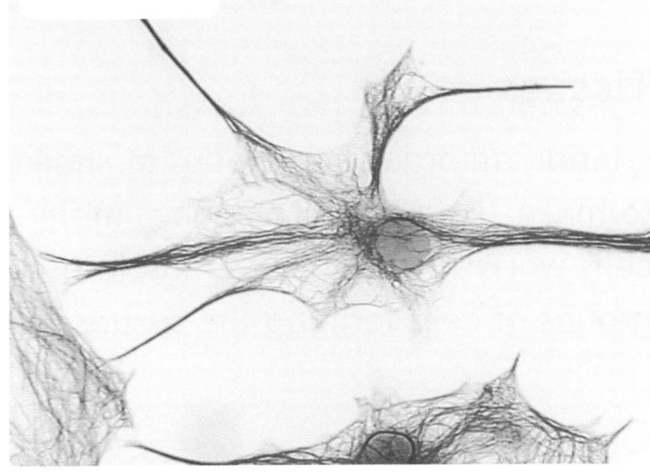

red and white blood cells

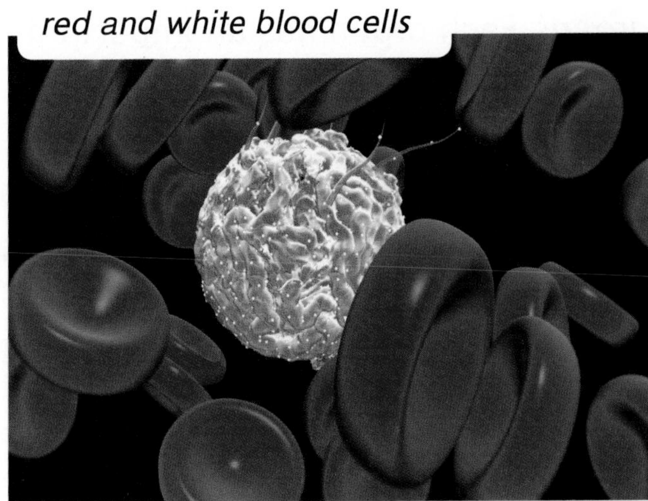

muscle cells

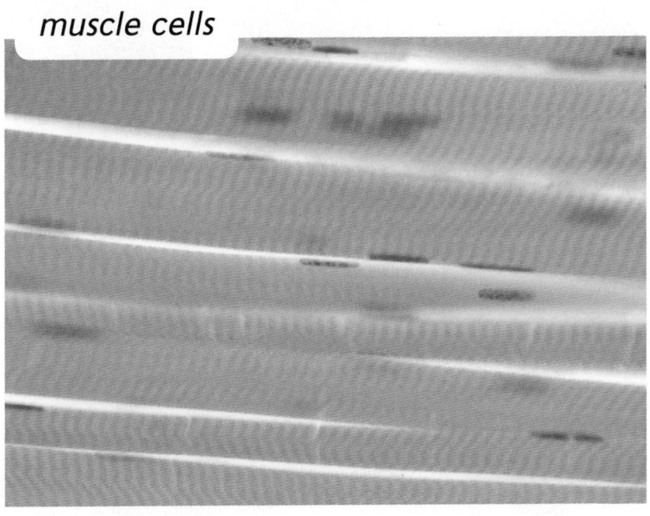

(t) Purestock/Getty Images; (c) MedicalRF.com; (b) Al Telser/McGraw-Hill Education

Cells Work Together

Tissues

Inside an organism, groups of similar cells join together to make *tissues*. Your body has many groups of muscle cells working together as muscle tissue. A flower has groups of cells forming the tissues of its leaves.

Organs

A living thing's tissues are organized together into *organs*. Each organ has an important job to do. The heart is the body's organ that pumps blood. A heart is mostly made of muscle tissue, but it also has nerve tissue. Signals from nerve tissue keep the heart pumping.

Levels of Organization

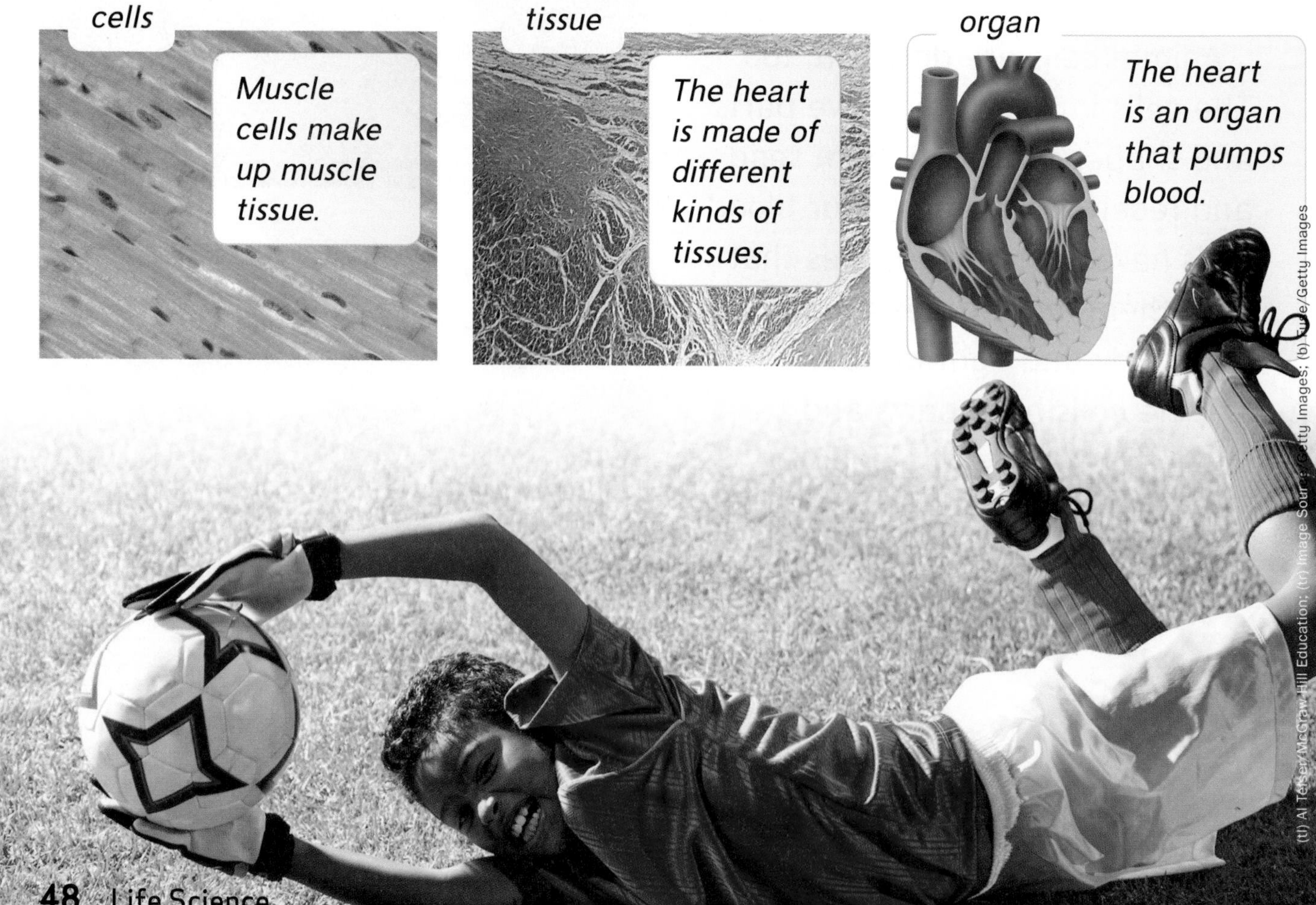

Organ Systems

Different organs join together to form *organ systems*. Organ systems make life possible. The heart works as part of the circulatory system, which moves the blood all through the body. The blood moves oxygen and nutrients to all of the body's cells and carries away wastes.

Every kind of living thing has its own set of organ systems. A tiger has a circulatory system that includes a heart. A mosquito does not. Organisms have the cells, tissues, organs, and organ systems they need to live and meet their needs.

Make Connections

Jump to the *Human Body* section to learn about your organ systems.

The circulatory system pumps blood to all the body's cells.

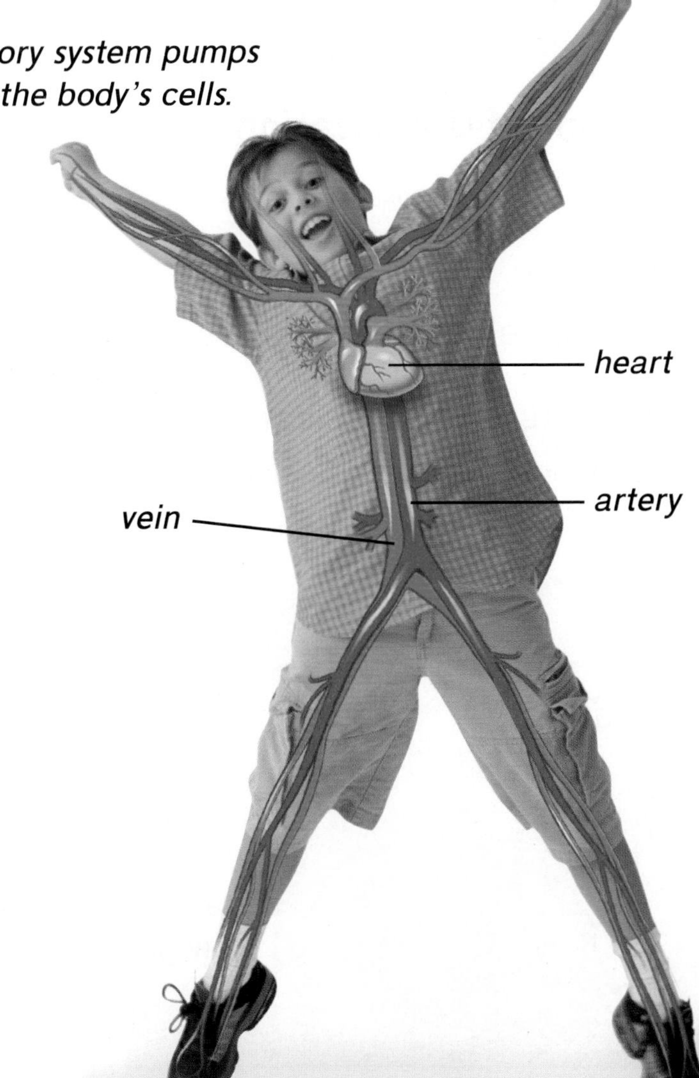

Types of Living Things

Classifying Living Things

Living things are called organisms. Some types of organisms, such as bacteria, are very small and have simple structures. Others, such as plants and animals, are larger and more complex.

Scientists classify living things by the distinctive characteristics, or **traits,** they share. When classifying an organism, scientists ask certain questions:

1. How does the organism get its food?
2. Does the organism move? If so, how does it move?
3. Does the organism have one cell or many cells?

Scientists have developed a system of classification for living things based on the answers to these questions.

This organism is a slime mold. Sometimes it looks like a colorful blob. At other times it looks more like a mushroom. These characteristics make it difficult for scientists to classify. Slime molds are now grouped with protists.

Matauw/iStock/Getty Images

Kingdoms of Living Things

Scientists organize living things into six *kingdoms*, or large groups, based on their shared characteristics. All the organisms in a kingdom share the same basic traits.

Until about 150 years ago, scientists classified all organisms as either plants or animals. Then scientists realized some organisms did not fit well in these kingdoms. Now plants and animals each have their own kingdom. The other four kingdoms are made up of organisms with simpler structures. Study the chart to learn the main characteristics of the organisms in the six kingdoms.

Did You Know?

Archaea are organisms that may live in extremely hot or salty environments. They may produce methane, which is also known as swamp gas.

The Six Kingdoms of Living Things

Kingdom	How many cells do these organisms have?	Can these organisms make their own food?	Do these organisms move from place to place?
Archaea	one	Most do not make their own food.	some
Bacteria	one	Some can make their own food.	some
Protists	one or many	Some can make their own food.	some
Fungi	one or many	no	no
Plants	many	yes	no
Animals	many	no	yes

Plants

Scientists estimate that there are about 400,000 different species of plants on Earth. Plants come in many shapes and sizes. Although plants can look very different, they share many of the same characteristics. Most plants have at least some parts that are green. Most grow in the ground. Plants cannot move around as animals can.

Plant Parts

Many plants have the same basic *structures*, or parts. Most have roots, stems, and leaves. These structures help plants get what they need to survive. Some plants have flowers, fruits, and seeds. Seeds can be found inside fruits. Nuts, like acorns, are also seeds. Other plants, such as pine trees, have cones that make seeds. Flowers, seeds, and cones help plants reproduce.

Roots The structures that hold a plant in the ground are called *roots*. Roots absorb water from the soil. In the process of absorbing water, roots also take in nutrients. *Nutrients* are materials that living things need for life processes. Some roots store food for a plant. These roots can provide food

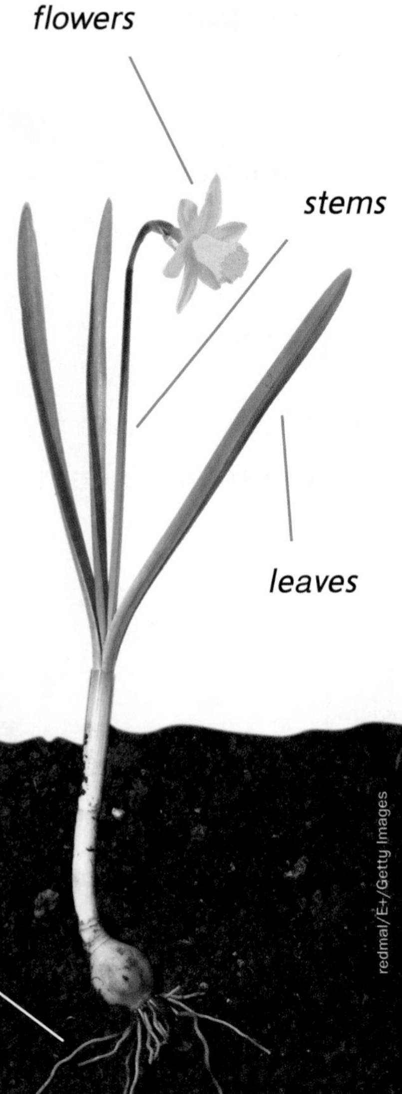

flowers

stems

leaves

roots

redmal/E+/Getty Images

for people, too. Sweet potatoes, radishes, and carrots are roots that you can eat.

Stems The structures that hold up a plant are called *stems*. Stems carry water and nutrients from the roots of the plant to the leaves through tubes. The leaves of a plant attach to the stem. Some stems also store nutrients and water. Some are soft and green, like the stem of a daisy. Other stems are thick and strong. Tree trunks are stems that are usually tall, hard, and woody. Shrubs may also have woody stems like those of trees.

Leaves Plants do not eat food like you do. Instead, they use energy from sunlight to make their own food. Most plants make food inside their leaves. Leaves also take in gases from the air and allow other gases to escape. Leaves come in many shapes and sizes. Some leaves are wide and flat, while others are needle-shaped.

Flowers Many plants have flowers. Flowers help the plant reproduce. When conditions are right, flowers produce fruits, which contain seeds.

What Could I Be?

Plant Biologist

Interested in learning more about plants and how they interact with the world around them? A plant biologist studies everything about plants. Some even use biotechnology to develop new and improved plants! Learn more about a career as a plant biologist in the Careers section.

Chris Sattlberger/Cultura RF/Getty Images

Needs of Plants

Like all organisms, plants have needs. Plants get everything they need from their environment. Living things that do not get what they need may die. Plants need carbon dioxide, water, light, nutrients, and space to live.

Carbon Dioxide Plants get a gas called carbon dioxide from their environment. Plants take in the gas through their leaves. They use this gas to make food. Oxygen is another kind of gas. It is made as plants make food. Plants give off the oxygen they make through their leaves. Animals breathe in this oxygen to stay alive. Plants also need oxygen.

Water Like all living things, plants need water. They take in water through their roots. Water travels from the roots, up the stem, to the leaves. Plants use water for many life functions. Water helps a plant stand up. It keeps a plant from wilting. A plant also uses water to make food.

Too much water can be harmful to plants. The plant on the right had too much water.

David Cook/blueshiftstudios/Alamy

Light There is a reason you do not see plants growing in caves. There is no light! The green parts of plants collect the energy in light and use it to make their own food.

Nutrients In addition to light, air, and water, plants need nutrients to survive. Nutrients are substances that help living things grow and stay healthy. Nutrients are dissolved in water. Plants absorb nutrients when they take in water through their roots.

Space Plants need space to grow and to get water and sunlight. Different plants need different amounts of space. Grass plants grow very close together. The plants grown in a vegetable garden do well when they have more room.

Fact Checker

Although most plants grow in soil, a plant can also grow in water without soil if it gets the nutrients it needs.

Sunflowers need a lot of light to grow.

(l) David R. Frazier Photolibrary, Inc./Alamy; (r) Pixtal/age fotostock

Reproducing with Flowers

How a plant germinates, grows, and reproduces is its **life cycle**. A cherry tree makes new seeds in its flowers. Inside the flowers, male parts make pollen, and female parts make eggs. Bees visit the flowers. They carry pollen from one flower to another. Pollination occurs when pollen and an egg meet. Pollination causes seeds to form.

Cherries are the fruit that forms around these seeds. The fruit protects the seeds as they develop. After the fruit ripens, it may fall onto the ground and break open. That releases the seed. Sometimes an animal eats the fruit and deposits the seed in its droppings. When soil and water conditions are right, the seed can germinate. A new cherry tree will grow.

What Could I Be?

Horticulturist

Thousands of plants are grown for food, fibers, spices, and medicines world wide. Horticulturalists study ways to support pollination of these plants. Learn more in the Careers section.

Life Cycle of a Cherry Tree

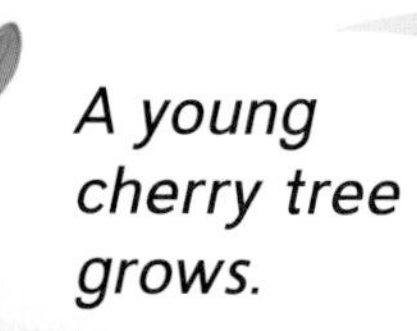

A young cherry tree grows.

Bees carry pollen to eggs in an adult tree. Seeds form.

Fruit forms around seeds.

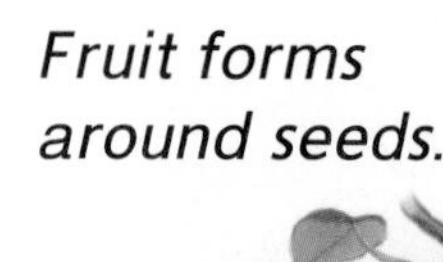

Ripe fruit falls and releases seeds.

A cherry seed germinates in the soil.

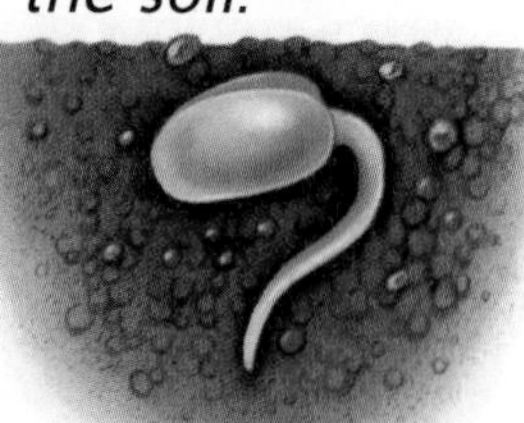

Photo by Stephen Ausmus, USDA-ARS

Reproducing with Cones

A pine tree's life cycle is similar to a flowering plant's, but pine trees don't have flowers. Instead, they form seeds inside cones. Pollination takes place when the wind blows pollen from male cones into female cones. Seeds deveolp in the female cones. When the seeds are ready, the cone opens, and wind blows the seeds to new places.

Plants that produce cones for reproduction are called *conifers.* Conifers have needle-shaped leaves. Most conifers keep their needles year-round. Those plants are often called evergreens. However, not all conifers are evergreens. For example, the Western larch's needles turn yellow and drop off during autumn.

Did You Know?

Lodgepole pine cones stay closed until the great heat of a forest fire pops them open. Their seeds have space to grow in the open land the fire leaves behind.

Skill Builder

Read a Diagram
Read the labels and follow the arrows in the diagram to understand a pine tree's life cycle.

Life Cycles of a Pine Tree

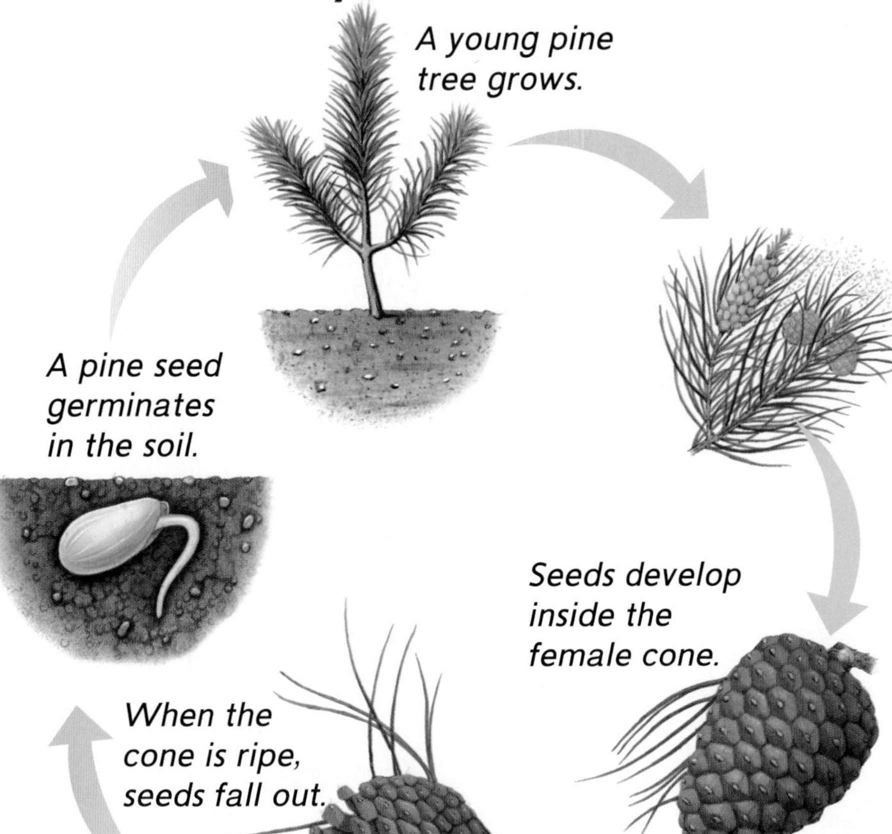

From Seed to Plant

If you have ever eaten peas, corn, or nuts, you have eaten seeds. Inside each seed is an *embryo*, a young plant ready to start growing. Also inside the seed is food for the embryo to use as it grows. The outside of a seed is tough, to protect the embryo.

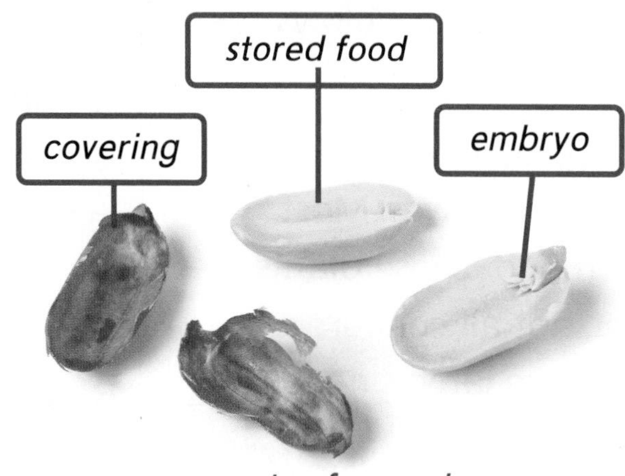

parts of a seed

When conditions are right, a seed will germinate. To *germinate* means to start growing. It waits until the temperature is right and the soil is wet enough. Some seeds wait for months or years before germinating.

A germinating embryo soaks up water until it swells. Then it bursts through the seed's outer covering. Over time the seed grows to be a mature plant.

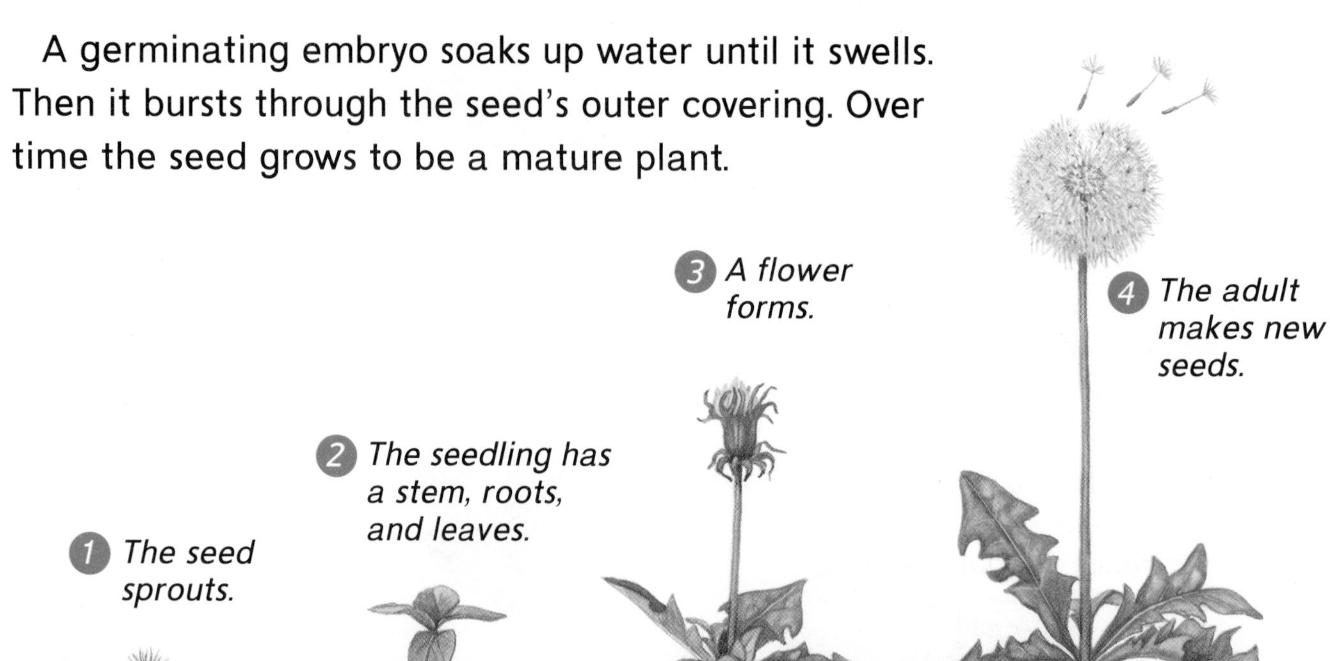

Ken Cavanagh/McGraw-Hill Education

Reproducing Without Seeds

Some plants reproduce without making seeds. For example, ferns reproduce by making spores. Spores are like seeds in some ways. They fall on the ground or blow in the wind. They can become a new fern plant. Unlike a seed, a spore does not have food for its embryo.

New stems and leaves can grow from potato "eyes."

New plants can also grow from parts of plants. The part of a potato plant that we eat is a tuber. A tuber is a large underground stem. We call the small white buds on potatoes "eyes." New stems and leaves can grow from these buds. Other plants, such as onions, can grow from an underground stem called a bulb. A new basil plant can grow from a stem or a leaf that is placed in water.

Life Cycle of a Fern

Adult ferns grow and release spores.

A spore grows into a small organism with male and female parts.

A young fern grows when cells from the male and female parts join.

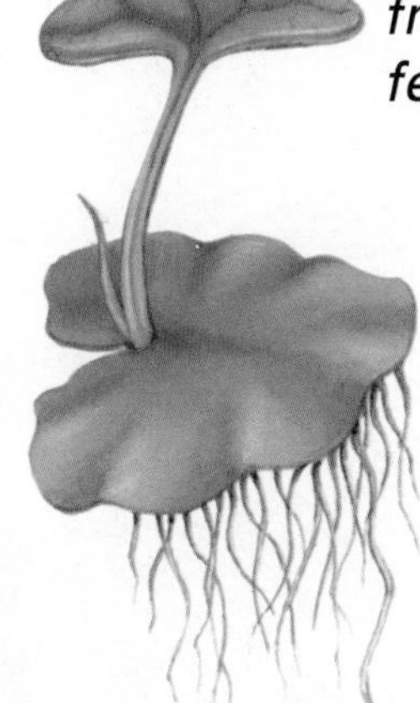

The Importance of Leaves

There are many different kinds of plant leaves. Maple leaves are broad and flat. Pine needles are thin and pointed. No matter how they are shaped, leaves all have the same job. Leaves make food for the plant.

To make food, plants need a gas from the air, water, and energy from sunlight. To capture the energy of sunlight, leaf cells contain a substance called *chlorophyll.* Chlorophyll is what makes leaves green. The underside of every leaf has tiny holes. These holes let in carbon dioxide, a gas that is part of the air around us. Most plants get water from the soil. Water enters a plant through its roots. The water travels up through the stem of the plant to the leaves.

maple leaf

pine needles

This photo shows one of the many microscopic holes in leaves.

Leaves are like little factories. Using carbon dioxide, water, and energy from the sunlight, they make food for the plant through a process called *photosynthesis*. During photosynthesis, plants use energy from the Sun to change carbon dioxide and water into sugar and oxygen. The sugar is the plant's food. Sugars give plants the energy they need to live and grow. When other organisms eat plants, they get the energy that the plant has made. Plants release the oxygen back into the air. Most organisms need oxygen to live. You inhale oxygen every time you breathe.

Skill Builder

Read a Diagram

Follow the arrows. Name the things that go into the plant. Then name the things that flow out of the plant.

Photosynthesis

***Sunlight** soaks into leaves and provides energy.*

***Oxygen** flows from the leaves as plants make food.*

***Carbon dioxide** flows into holes in leaves.*

***Food** made inside leaves travels to the rest of the plant.*

***Water and nutrients** flow from the soil into the roots, and from the roots to the leaves.*

Animals

Animal Needs

In order to stay alive, animals need certain things. These include food, water, oxygen, space, and shelter.

Food Animals need food because it gives them energy to move and grow. Different animals have different parts to help them get food. Some meat eaters, like tigers, have sharp teeth to help them bite and tear meat. Many plant eaters, such as cows, have large, flat teeth for chewing. Birds have sharp beaks for tearing meat or breaking into seeds. Animals that do not have teeth or beaks use different methods. For example, houseflies spit saliva on their food, wait until the food becomes gooey, and then slurp it up.

Did You Know?

More than half the human body is made of water!

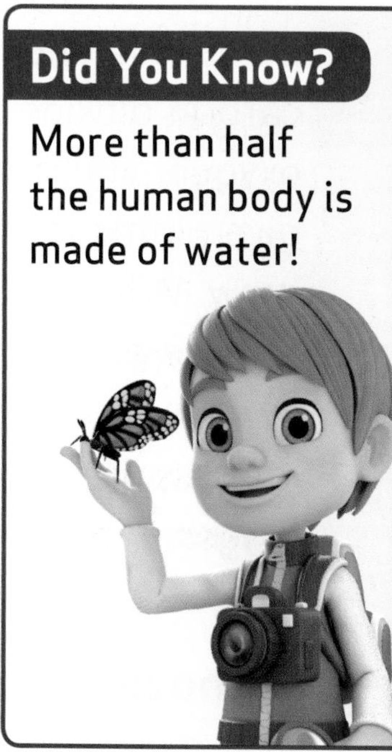

Elephants use their trunks to lift drinking water to their mouths.

Water Animals need water because it helps them turn food into energy and get rid of waste. Animals get water in different ways. Dogs lap up water with their tongues. Bees slurp water. Gila monsters get water from food.

Oxygen Animals need oxygen, which is a gas. Animals get oxygen by breathing. Most land animals use lungs to get oxygen from the air. Some animals that live in water get oxygen from the water by using gills. Animals with no lungs or gills get oxygen in other ways. For example, worms breathe through their skin.

Space Animals need space to move around, grow, find food, and raise their young. Different animals need different amounts of space. A salmon can swim hundreds of miles, but a gopher may live in the same field for its whole life.

Shelter Animals need shelter, or a safe place to be. No animal can be alert all the time, which means it needs somewhere safe to go. Zebras live in herds, so some zebras can keep watch while others sleep. Turtles have hard shells to protect them. Ferrets dig tunnels in the ground, where they can hide.

As this shark swims, water moves in its mouth and out its gills. The gills take oxygen from the water, letting the shark breathe.

When in danger, a turtle will hide its head, legs, and tail inside its shell.

Animal Groups

There are millions of types of animals on Earth. To make them easier to understand and study, scientists put animals into groups. These groups are based on shared traits. One trait scientists use is whether or not an animal has a backbone.

A backbone is a line of many small bones that runs down the middle of an animal's back. An animal with a backbone is called a *vertebrate*. Humans are vertebrates, and so are dogs, fish, pigeons, and iguanas.

An *invertebrate* is an animal that not only has no backbone—it does not have any bones at all! Invertebrates are more common than vertebrates. Jellyfish, beetles, snails, and spiders are invertebrates.

Word Study

Each bone in the backbone is called a *vertebra*. The plural is *vertebrae*. That's why animals with backbones are called vertebrates.

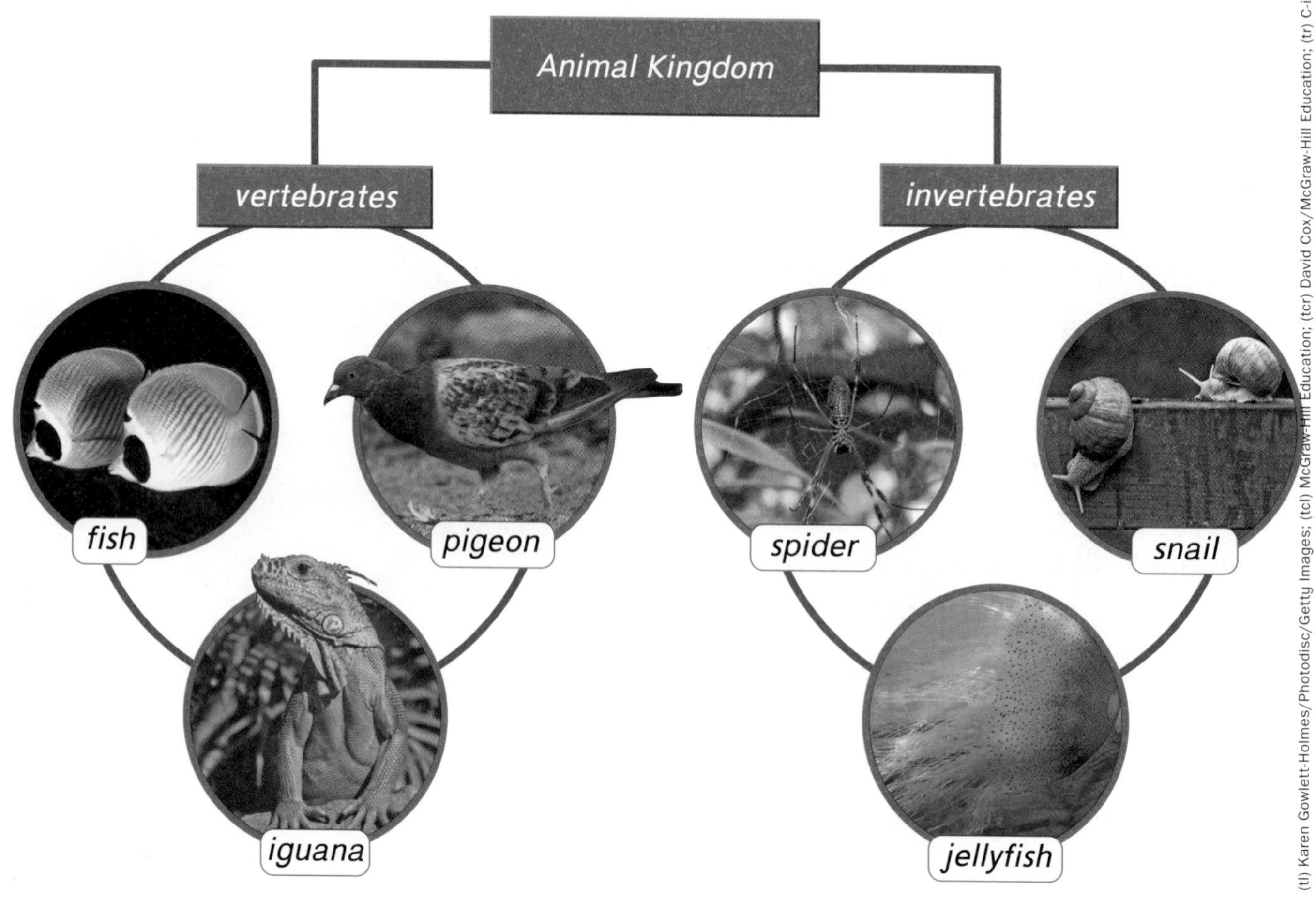

(tl) Karen Gowlett-Holmes/Photodisc/Getty Images; (tcl) McGraw-Hill Education; (tcr) David Cox/McGraw-Hill Education; (tr) C-images/Alamy; (bl) Kevin Schafer/Alamy; (br) Purestock/SuperStock

Invertebrates

Invertebrates can be found all over Earth. Most are small, like insects. A few, such as the giant squid, can grow as long as a school bus.

Instead of bones, invertebrates have other structures that hold up and protect their bodies. Many have a thin, hard covering called an *exoskeleton*. Lobsters, grasshoppers, and spiders have exoskeletons. Some, such as snails and clams, have hard outer shells. Others, such as seas stars, have shells inside their bodies! A few, such as jellies, depend on the water in which they live to support their bodies.

Arthropods make up the biggest group of invertebrates. There are over 800,000 named species of arthropods. Arthropods have exoskeletons, bodies with more than one segment, and legs with joints. Spiders, ants, centipedes, and shrimp are all arthropods. The photos below show some of the many kinds of arthropods.

Word Study

In Greek, *exo* means "outer." In other words, an exoskeleton is a skeleton on the outside of the body. In order to grow, animals with exoskeletons must shed their outer layer, or molt.

The joints in this spider's legs help it run quickly. They also let the spider use its legs to grab things or climb up a single strand of web.

Centipedes have one pair of jointed legs on each body section.

Hermit crabs use found shells as extra protection, especially when they are molting. As hermit crabs grow, they have to keep finding bigger shells to move into.

Vertebrates

Vertebrates are the animals you are probably most familiar with. Scientists organize vertebrates into five groups: mammals, birds, reptiles, amphibians, and fish.

Mammals

Mammals are a group of vertebrates that have fur or hair. Mice, dogs, elephants, and people are all mammals. Most mammals are born live, meaning not hatched from an egg. Female mammals make milk to feed their young. They care for their young until the young can find food on their own. All mammals have lungs, not gills. Mammals that live in the water, like whales and dolphins, swim to the surface to breathe air.

Mammals are warm-blooded. This means that their body temperature does not change much. Being warm-blooded allows mammals to live in many places. Polar bears live in the cold arctic. Tigers live in swamps, grasslands, and rain forests.

A mammal's first food is milk from its mother.

Dolphins breathe air into their lungs through a hole in the top of their heads.

Birds

Birds are another familiar group of vertebrates. Birds have feathers, two wings, and two legs. Birds are built to fly. Most birds have hollow bones, which make them very light. Like mammals, birds use lungs to breathe air and are warm-blooded. Unlike mammals, all birds lay eggs. Most birds lay their eggs in a nest and then keep the eggs warm until they hatch.

Although all birds have wings, not all birds can fly. Penguins use their wings as rudders and propellers to help them move easily through water. Ostriches use their wings to keep warm at night, when the temperature drops as much as 40°Celsius (100°Fahrenheit).

Some birds, such as ravens and parrots, are considered very intelligent. They can even learn to speak.

Most ducks feed and care for their young until the young can find food on their own.

Vertebrates - Reptiles

Reptiles are the vertebrates most closely related to birds. Like birds, they have lungs. Also like birds, most reptiles lay eggs. However, some, like the boa constrictor, give birth to live young. Reptiles do not have wings or feathers, and they cannot fly. Almost all reptiles are cold-blooded. Their body temperature changes with the temperature of the environment.

Some reptiles live on land and some live in water. Many reptiles, such as crocodiles, are comfortable both on land and in water. They are comfortable because reptiles are covered in waterproof scales that keep their bodies from drying out.

Common reptiles include turtles, crocodiles and alligators, snakes, and lizards.

Fact Checker

Some people think that all snakes are venomous. In fact, only about 10 percent of snake species produce venom. Of the 100 snake species native to the United States, only 20 produce venom.

Many reptiles have four legs but some, such as snakes, do not have any!

At over 70 kilograms (154 pounds), the alligator snapping turtle is the largest freshwater turtle.

(l) Jason Ondreicka/Alamy; (r) Gary M. Stolz/U.S. Fish and Wildlife Service

Vertebrates - Amphibians

Amphibians are vertebrates that live part of their lives in water and part on land. Most amphibians start life underwater. A young amphibian that lives in the water is called a larva or tadpole. Larvae have gills and look like fish. As they get older, amphibians go through metamorphosis, which means their body form changes. Some amphibians, such as newts, grow legs. Others, such as caecilians, look like worms or snakes.

When they live on land, amphibians breathe air. Most, but not all, amphibians grow lungs. Some amphibians can also breathe through their skins. Like most reptiles, amphibians are cold-blooded. Unlike reptiles, amphibians do not have scales.

Frogs, toads, salamanders, caecilians, and newts are all amphibians.

Some adult amphibians, like this frog, breathe through lungs or their skin.

A salamander is an amphibian that lives near water or other cool, damp places.

This caecilian is an amphibian that lives mostly in the ground.

Make Connections

Jump to *Animal Life Cycles* to learn about amphibian metamorphosis.

Vertebrates - Fish

Fish are the last group of vertebrates. All fish have gills and spend their entire lives in water. There are about 32,000 different species of fish. Fish live almost everywhere there is water. They live in freshwater and saltwater, in streams and oceans, and in shallow water and deep water. Most fish are covered in scales, feel slimy, and lay eggs. Some sharks give birth to live young.

Almost all fish are cold-blooded. Some fish, however, are able to maintain their body temperatures like warm-blooded organisms.

Different fish grow to different sizes and eat different foods. Eels can grow to be ten feet long, but some carp are no bigger than your pinky fingernail. Sharks eat only meat, but rainbowfish eat plants. Other fish strain tiny floating organisms from the water for food.

Lamprey eels have no jaws. They have a mouth like a straw with teeth inside to suck up prey or to suck blood!

A flat shape and smooth skin help fish such as this stingray move through the water.

Animal Life Cycles

Metamorphosis

When many animals begin their lives, they look like smaller versions of their parents. Puppies, for example, look like small dogs, and foals look like small horses. Other animals look different at different stages of life. Some animals change shape through a process called **metamorphosis**. Most insects and amphibians undergo metamorphosis.

Life Cycles - Frogs

The frog is an amphibian that undergoes metamorphosis. Follow the arrows in the diagram below to see the frog's life stages.

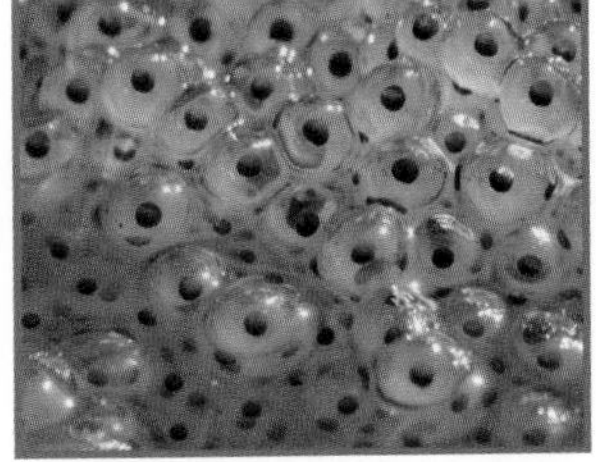

Egg Frogs lay eggs in water.

Life Cycle of a Frog

Tadpole Young frogs, or tadpoles, hatch. Like fish, they swim and breathe with gills.

Becoming an Adult A tadpole starts to grow legs and lungs.

Adult Now the frog looks like its parents. It moves onto land and can reproduce.

Life Cycles - Insects

Like amphibians, insects undergo metamorphosis. Insects start life as eggs. The egg contains food that the young animal needs. When the young animal has grown enough, it hatches, or breaks out of the egg. Then it goes through several stages. Eventually, it grows into an adult that can have its own young.

Life Cycle of a Ladybug

Egg A ladybug starts life as an egg.

Larva When an insect hatches, it is called a larva. A ladybug larva eats bugs and grows.

Pupa The larva changes into a pupa. It forms a hard shell. Inside, it grows wings.

Adult The adult ladybug has red wings. Females can lay eggs

(t) Andrew Palmer/Alamy; (cr) Inzyx/iStock/Getty Images; (cl) Photowood Inc./Alamy; (b) Sheryl Salisbury/Moment Open/Getty Images

Life Cycles - Reptiles, Fish, and Birds

Most reptiles, fish, and birds lay eggs. When these eggs hatch, the babies look like small versions of their parents. This means that they do not undergo metamorphosis. Reptiles, fish, and birds have similar life cycles, but there are some differences. Birds lay eggs in nests, fish lay eggs in water, and reptiles lay eggs on land. Most birds look after their young, but most fish and reptiles do not.

A good example of the life cycles that reptiles, fish, and birds go through is the turtle. Look closely at the stages shown in the diagram below.

Life Cycle of a Turtle

Egg Females crawl to the beach to lay eggs in the sand.

Young Sea turtles hatch on the beach and quickly crawl to the ocean.

Adult Turtles grow to 140 kilograms (300 pounds). Females stay in the sea until they are ready to lay eggs.

Life Cycles - Mammals

Like reptiles, fish, and birds, young mammals look a lot like their parents. This similarity means they do not undergo metamorphosis. Most mammals are born live. They do not hatch from eggs.

Many young mammals are weak and slow. They cannot find their own food. They need their parents in order to survive. Mammal parents teach their young important lessons. For example, lion parents teach their cubs how to hunt. Deer parents teach their fawns to run away from danger.

Skill Builder

Read a Diagram

Follow the arrows and read the captions to understand the cheetah's life cycle.

Life Cycle of a Cheetah

Cub Most female cheetahs have three to five cubs at once. They protect and feed the cubs.

Young Cheetahs learn and practice the skills they will need to hunt.

Adult Cheetahs grow big and can reproduce. Adults are as fast as a car on a highway.

Animal Reproduction

The life cycle of every animal includes **reproduction**. This is when parent animals make offspring.

One Parent

Some simple invertebrates reproduce by budding. A bud forms on the adult's body. After the bud breaks off, it grows into an adult. Sea stars can reproduce by regeneration. This process occurs when a whole animal develops from just a part of the original animal. Both budding and regeneration produce copies of the parent. These copies have exactly the same characteristics, or traits, as the parents.

Two Parents

Another kind of reproduction requires cells from both a male and a female parent. When the female and male cells join, offspring are produced. This new offspring has traits from both its parents. It is not identical to either parent. Fish, amphibians, reptiles, birds, and mammals reproduce with two parents.

Like all mammals, giraffes reproduce with two parents.

Inherited Traits

Every organism has traits that make it unique. Eye and hair color are examples of traits. The shape of your earlobes and whether you can roll your tongue are traits. Traits help you recognize and describe an organism.

Where do an organism's traits come from? Part of the answer is heredity. *Heredity* is the passing on of traits from parents to offspring. Traits that come from parents are called **inherited traits**. Your eye color and hair color are inherited traits. The number of legs an animal has is also inherited. Inherited traits make animals look like their parents.

Inherited traits can also affect the way an animal acts. Reflexes, such as blinking, are actions that parents pass on to their offspring. So are instincts. An **instinct** is a way of acting that an animal does not have to learn. Birds build nests and spiders spin webs by instinct. Many instincts help animals survive.

What Could I Be?

Geneticist

Geneticists study traits and heredity in organisms. They develop ways to develop desired inherited traits in plants and animals. Learn more about a career as a geneticist in the Careers section.

poodle

labradoodle

yellow labrador

The labradoodle has one poodle parent and one yellow Labrador parent. Can you identify which traits came from which parent?

Learned and Environmental Traits

Some of your traits come from your parents. Others you have learned. People and animals can gain new skills over time. These new skills are called *learned traits*. Learning to ride a bicycle or play the piano are learned traits.

Some traits are affected by the environment. For example, your hair may get lighter from being in sunlight. A rabbit may grow fat when it finds a lot of food. It may grow thin when food is hard to find. These types of traits are *environmental traits*.

Learned traits and environmental traits are not passed from parents to offspring. Your parents may know how to ride a bike, but you still had to learn that skill yourself. If your parent has a scar, you were not born with the same scar.

Did You Know?

If you have a kitten or puppy around an older cat or dog, the older pet will often teach the younger pet lessons it learned from humans. For example, adult cats may train kittens to use litter boxes, and dogs may help teach puppies to walk by their owner's side.

Both the riders and the horses are demonstrating learned traits. The people have learned how to ride horses. The horses have learned how to be ridden.

Image Source

Ecology

Ecosystems

Living things depend on one another. They also depend on nonliving things like sunlight. Living and nonliving things that interact in an environment make up an **ecosystem**.

Earth has many different kinds of ecosystems. Some are on dry land, and others are underwater. Some are warm, and some are cold. Some have many different kinds of organisms, while others have only a few. An ecosystem may be a pond, a swamp, or a field. It may be a meadow, a river, or an island. An ecosystem may be as small as a puddle or as big as an ocean.

A Pond Ecosystem

Crane flies eat plants and algae. They lay eggs in water.

Different organisms live in different parts of an ecosystem. Fish live in the water. Water is their *habitat*, or home. A cattail's habitat is along the edge of a pond. An insect's habitat may be on a cattail. Living things get food, water, and shelter from their habitats. Many different habitats make up an ecosystem.

Skill Builder

Read a Diagram

Looking carefully at how living things interact with one another and with nonliving things will help you understand the parts of an ecosystem.

Turtles climb out of the water to warm up in the sunlight.

Cattails grow well in wet soil. Animals use them as food and shelter.

Pond snails slide along the bottom, looking for plants and algae to eat.

Parts of an Ecosystem

All ecosystems are made up of living and nonliving things. Living things include plants and animals, as well as bacteria and algae. Frogs, birds, and fish are some living things in a pond. Plants, such as water lilies, grasses, and cattails, are also living things. Nonliving things include rocks, minerals, soil, water, and air. Sunlight, water, and rocks are examples of nonliving things in a pond.

A **resource** is a material that living things use to survive. Living things get resources from their ecosystems. For example, plants and animals get the air and water they need. Plants use sunlight to make food. Animals use plants and other animals as food.

The water, sunlight, and soil are nonliving things in this pond ecosystem.

Melissa King/Cutcaster

Parts of an Ecosystem

Living	Nonliving
plants	rocks
animals	minerals
bacteria	soil
algae	water
	air
	sunlight

Skill Builder **Read a Table**

Read the column heads in a table to find out what types of things appear in each column. The column heads of this table tell you that only living things are listed in the left column. Only nonliving things are listed in the right column.

Interactions in Ecosystems

The different parts of an ecosystem interact. The parts depend on and affect one another. Living things depend on other living things. For example, animals depend on plants for food and shelter. Squirrels depend on trees for acorns to eat and gather branches to make nests in trees. The squirrel can change and affect how trees grow. Trees and other plants also give off air that animals need in order to breathe.

Living things also depend on and affect nonliving things in an ecosystem. The grass in a meadow could not grow without air, water, sunlight, and soil. *Soil*, a nonliving thing, is made up of bits of rock and humus. *Humus* is broken-down plant and animal material. It contains nutrients and soaks up rainwater. Soil that is rich in humus holds plenty of water and nutrients that plants need. When chipmunks and rabbits dig in the ground, they break up rocks, which helps form soil.

Bison and other animals in this meadow depend on grasses for food. The grasses need air, water, sunlight, and soil to grow.

Animals, including humans, affect nonliving parts of their ecosystem when they dig in the ground and when they use water.

Forest Ecosystems

A *forest* is an ecosystem that has many trees. Different types of forests can be found in different parts of the world.

Toucans eat fruit from trees in this tropical rain forest.

Tropical Rainforest A *tropical rain forest* is a forest that is hot and damp. Rain forests have more types of living things than any other land ecosystem. Brightly colored birds live in trees along with mammals, insects, and reptiles.

Tropical rain forests are warm all year long. They get a lot of rain, but their soil is not rich in nutrients. Rain forest plants quickly absorb any nutrients in the soil.

Temperate Forest A *temperate forest* changes with the seasons. It is cold and dry in winter and warm and wet in summer. Temperate forests receive less rain than tropical rain forests. However, enough rain falls for large trees to grow. The soil is rich in nutrients and soaks up plenty of water.

Bears, foxes, squirrels, and deer find homes in temperate forests.

Desert Ecosystems

A *desert* is an ecosystem that has a dry climate. Fewer than 25 centimeters (10 inches) of rain fall in a desert each year. The Sonoran Desert is one of the largest deserts in North America. It is located in the southwestern United States, in Arizona, California, and northwestern Mexico.

Temperatures in most deserts vary widely between day and night. During the day, the Sun's heat warms the land and air. After sunset, the temperature drops quickly.

The soil in a desert is mostly sand. There is little humus to soak up rainwater. Rainwater trickles down through the sand. It goes deeper than most plant roots can reach.

Fewer plants and animals can survive in a desert ecosystem than in forests. Desert soil has little water and nutrients for many plants. Desert plants that do survive usually grow far apart. Most desert animals find shade in which to rest during the heat of the day. They hunt at night when temperatures are cooler. Jackrabbits, rattlesnakes, and lizards are common desert animals.

Fact Checker

Not all deserts are hot. Some deserts, such as Antarctica, are dry and cold.

The saguaro cactus is one kind of plant found in the Sonoran Desert.

This collared lizard hunts insects and other lizards in the Sonoran Desert.

Ocean Ecosystems

An *ocean* is a large body of salt water. Earth has five oceans, which are all connected. They are the Atlantic, Pacific, Indian, Arctic, and Southern oceans. The Pacific Ocean is the largest. It covers about one-third of the planet.

Billions of living things are found in the five oceans, but almost all ocean organisms live in the shallow waters that are less than 100 meters (330 feet) deep. Here the water is lit and warmed by the Sun. Green plants and algae get enough sunlight to grow. They attract animals that depend on them for food and shelter. Few creatures can survive in cold, dark ocean depths, which can be more than 1,500 m (5,000 ft) deep.

This fangtooth fish is one of a few organisms that live in the icy, dark waters of the deep ocean.

What Could I Be?

Marine Biologist

Want to study sharks, observe whales, or learn how coral reefs form? Marine biologists study ocean organisms. Some work on boats, some dive underwater, and some do research in labs on land. Learn more about a career as a marine biologist in the Careers section.

Tiny animals called coral form this reef in the shallow part of a tropical ocean.

Wetlands

A *wetland* is an ecosystem in which water covers the soil for part or all of the year. Wetlands are often found along the edges of rivers, lakes, ponds, and oceans. They contain freshwater or salt water.

In the past, people thought of wetlands as swamps to be drained, but wetlands are important ecosystems. Wetlands are habitats for many birds, amphibians, and insects. Many water organisms use wetlands as places to lay their eggs. Wetland soils are usually full of minerals that help plants grow. Wetlands help prevent land from flooding by collecting water. Wetland plants even help clean dirty water.

The land of this ecosystem is wet, but it may dry up for part of the year.

Wading birds, such as this Great Blue Heron, meet their needs in wetland ecosystems.

Energy in Ecosystems

All living things in an ecosystem need energy. They make or take in food to get energy. An organism's role in its ecosystem depends on how it gets energy. There are three main roles in ecosystems.

Producers A **producer** is an organism that makes its own food. Green plants and algae are two examples. Most producers use energy from the Sun to make their own food.

Consumers A **consumer** is an organism that eats other organisms. It gets energy from the organisms it eats. All animals are consumers, including insects, fish, and horses. Humans are consumers too.

Decomposers A **decomposer** is an organism that breaks down dead plant and animal material. As they feed on this material, decomposers release nutrients into the water or soil. These nutrients help plants and other organisms. Bacteria and fungi are two common decomposers.

Fuse/Getty Images

Consumers

Scientists group consumers by the kinds of food they eat. There are three main kinds of consumers.

herbivore

Herbivores Organisms that eat only plants are *herbivores*. Cows and squirrels are herbivores. So are deer, elephants, rabbits, and grasshoppers.

carnivore

Carnivores Some consumers, such as herons, eat other animals. Animals that eat only other animals are *carnivores*. Hawks, seals, and cats are carnivores. So are wolves, sharks, and lions.

omnivore

Omnivores Animals that eat both plants and animals are *omnivores*. Raccoons, crows, blue jays, and bears are all omnivores. Many humans are omnivores too.

Did You Know?

Herbivore honeybees help plants reproduce when they take nectar from flowers for energy. The only honeybees you will likely ever see are female worker honeybees.

Food Chains

All organisms in an ecosystem need energy from food to live and grow. Most are a source of energy for other organisms as well. They pass on energy to organisms that eat them. A *food chain* shows how energy passes from one organism to another in an ecosystem.

The first organism in a food chain is always a producer. Producers make their own food using the energy in sunlight, water, and carbon dioxide. For this reason, the energy in most food chains starts with the Sun.

Notice that the first thing in the food chain is the Sun. The Sun is followed by sedge grass and algae. These producers get energy from the Sun and use it to make their own food.

A Pond Food Chain

Since consumers eat producers, the next organisms in a food chain are always consumers. A food chain may have many consumers. In the first pond food chain, grasshoppers get energy by eating sedge grass. Turtles eat the grasshoppers to get energy. Bald eagles eat the turtles.

In the second pond food chain, pond snails get energy by eating algae. Sunfish eat the snails. Bass eat the sunfish. The snails, sunfish, and bass are all consumers.

Last in a food chain are decomposers. When the eagle and bass die, bacteria get energy from the dead animals.

Predators hunt other organisms for food. The organisms they hunt are *prey*. The bass is the prey of the heron and eagle.

Turtles eat grasshoppers.

Bald eagles eat turtles.

Sunfish eat snails.

Largemouth bass eat sunfish.

Food Webs

Most animals eat several kinds of food. They are part of several food chains. Several connected food chains form a *food web*.

The diagram shows a pond food web. Trace the arrows from the largemouth bass to the heron and bald eagle. They show that herons and eagles eat bass. The bass is part of more than one food chain.

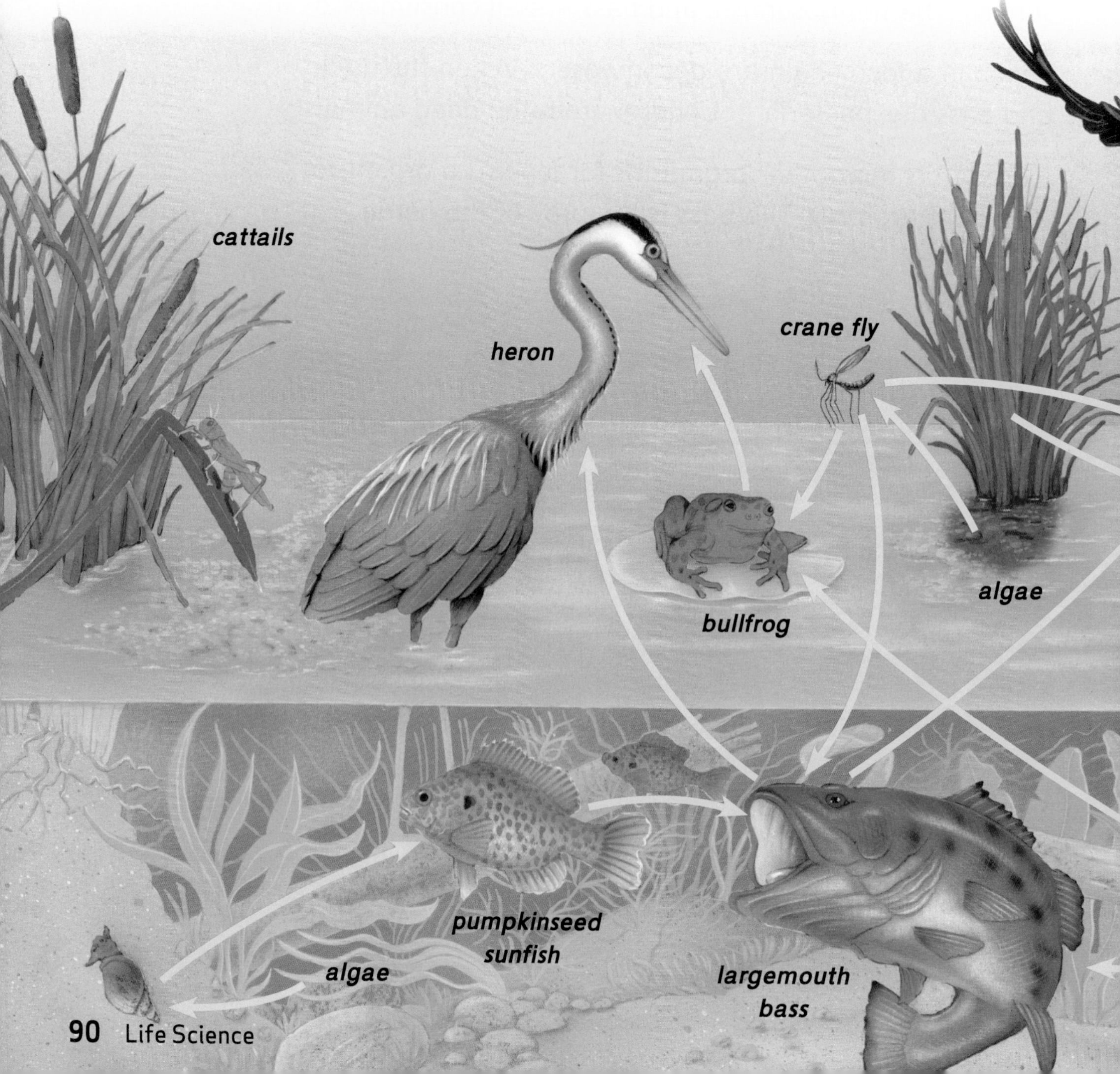

Food webs show how energy passes from one organism to another in an ecosystem. They show how all living things are connected and how organisms compete for food. For example, many animals eat crayfish. If snakes eat all the crayfish, other animals might go hungry.

Scientists predict how organisms affect each other. They look at what would happen if all the crane flies in a pond die. Bullfrogs and mallard ducks both eat crane flies. Without the flies, these predators might not have enough food. They might die, too.

Skill Builder

Read a Diagram

A food web diagram includes arrows. Arrows show the ways energy moves through an ecosystem from one living thing to another.

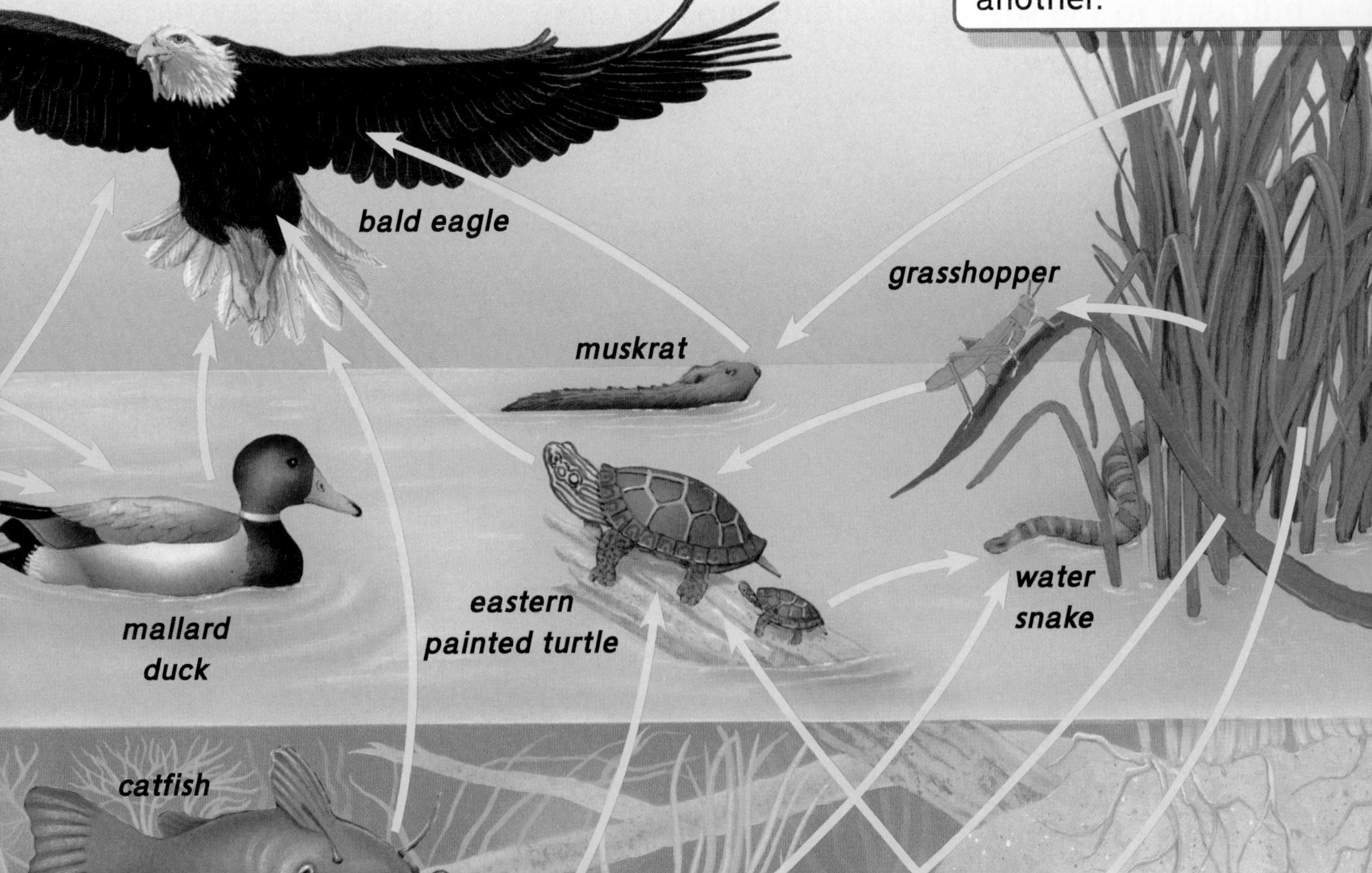

Changes in Ecosystems

Every living thing changes its ecosystem as it meets its needs. A spider spins a web to catch insects for food. A robin builds a nest for shelter. A cotton plant takes water from the soil. Organisms reproduce and grow in number. These events change an ecosystem in small ways.

Other living things make bigger changes to their ecosystems. For example, bacteria, worms, and mushrooms break down leaves and dead animals. These decomposers return valuable nutrients to the soil. Later, plants can use those nutrients to grow.

What Could I Be?

Ecoinformatics Specialist

Love data and numbers? Interested in how ecosystems are changing? Scientists who work in the field of ecoinformatics find ways of handling large amounts of data about ecosystems. Learn more about this career in the Careers section.

Seeds blow onto bare ground. The environment changes as plants take in water and nutrients.

As more plants grow, animals move to the environment. They use the plants for food and shelter.

People change ecosystems more than other organisms do. Some changes, such as planting trees, are helpful. Others, such as draining wetlands to build over them, harm ecosystems.

All the living things shown here are trying to secure resources that meet their needs and help them survive. But every ecosystem has a limited number of resources. As a result, living things must compete for them. **Competition** is the struggle among living things for resources. When organisms cannot compete, they cannot get the resources they need. They may die or move to another ecosystem.

In time the plants grow larger. They compete for water, space, and sunlight. Animals compete for food and water.

Trees block sunlight from reaching smaller plants. These plants may die as trees grow larger.

Adaptations and Behaviors

Adaptations

A frog sees an insect. The frog's long, sticky tongue springs out and catches the insect. The tongue curls back in, and the frog swallows its food whole. A frog's tongue is an example of adaptation. An **adaptation** is a structure or behavior that helps an organism survive in its environment.

Energy

Every living thing is adapted to get energy. The leaves of an elm tree capture sunlight to make food. Mountain lions have sharp teeth and claws to catch deer and other animals. The flat teeth of horses are perfect for chewing grass.

A polar bear has an adaptation called **camouflage**, which means it blends into its environment. Covered with white fur, the polar bear is hard to see on white snow and ice. That makes it easier for the bear to catch seals for food.

The frog uses its sticky tongue to capture an insect to eat.

Oktay Ortakcioglu/Vetta/Getty Images

Safety

Camouflage also helps living things stay safe. A snake's skin pattern may match the ground it lies on, making it hard for a predator to see. A turtle's shell is another adaptation for staying safe. When a turtle is in danger, it can pull its head, legs, and tail inside the shell.

Plants also have adaptations for staying safe. Holly plants have spines on the edges of their leaves. The spines stop many animals from eating the plant.

Did You Know?

A mimic octopus can change its movements and the arrangement of its parts to look like 15 different organisms, including sea snakes, brittle stars, giant crabs, and stingrays.

Protection from the Weather

Sea lions and walruses are adapted to living in cold climates. They have a layer of fat called blubber under their skin that helps them stay warm. An arctic fox grows a thick white coat when winter is coming. That coat keeps the fox warm even in very cold weather.

Some animals are adapted to survive in hot temperatures, such as in a desert. Camels have thick, leathery patches on their knees. The patches protect their legs so they are not burned when the camel kneels.

Because of its camouflage, this snake will be hard for a hawk flying overhead to see.

A camel has thick fur on its back that provides shade. Thinner fur on other parts of the body let heat escape.

Desert Adaptations

A desert is a very dry environment. It rarely rains in a desert. When rain does fall, it can pour down heavily. Temperatures are often very hot during the day and cold at night. Organisms that live in a desert have adaptations to help them survive in these conditions.

Skill Builder

Read a Diagram

The diagram shows two different ways desert plants get water from soil. Read the labels to understand the differences.

Water

Desert plants cannot depend on regular rain for their water. Instead, their roots are adapted to spread out widely or reach down deeply to find water. A desert plant has stems adapted for storing water. Some plants have thorns to keep away animals that might eat the stems or fruit. Many desert animals get their water by eating plants or other animals.

Temperature

Desert animals have adaptations to keep them from being too hot during the day. Coyotes and rattlesnakes are *nocturnal.* This means they are active at night, when the desert is cool. They sleep in the daytime when it is hot.

Jackrabbits stay cool by having thin bodies and long ears. This adaptation helps the heat escape from their bodies. Warm blood flows to a jackrabbit's large ears. Once there, heat is released into the air. Some animals have pale-colored bodies. Pale colors absorb less heat.

Snakes and lizards stay out of the blazing Sun by hiding in holes or under rocks. Vultures and eagles soar high above the hot desert sands. The air is cooler where they fly.

Warm blood flows through the jackrabbit's ears. Some of the heat escapes into the air.

This bat is nocturnal. It sleeps in the daytime when the desert is hot. At night, it eats nectar from cactus flowers.

Forest Adaptations

A forest is an environment in which many trees grow near one another. The tops of tall trees are in sunlight. Plants on the forest floor live in shade. Some organisms are adapted to life in the treetops. Others live on the ground.

Forest Plants

Tropical rain forests are very wet places. Too much water can harm plant leaves. Some leaves have grooves and drip-tips that allow water to drain off easily. Little sunlight reaches beneath the treetops. Plants living on the dim forest floor have large leaves to catch as much sunlight as possible.

These leaves have stopped making food. They drop off in fall. In spring, the tree will grow new leaves.

A temperate forest grows where winters are cold and dry. In winter there is less sunlight for plants to make food. Many trees have adapted by shedding their leaves each autumn. Without leaves, the trees need less water.

Rainwater drops off the tips of these leaves.

(t) Jim Powell/Alamy; (b) Christopher J. Bandera/Moment Open/Getty Images

Forest Animals

Animals have many different adaptations for survival in the forest.

Mimicry A stick insect looks so much like a stick that other animals don't notice it. **Mimicry** occurs when one living thing looks like another. Mimicry gives an animal a way to be hidden when not moving. It can help an organism hunt without being seen.

Defense When a predator comes after a skunk, the skunk sprays a stinky chemical at it. A porcupine defends itself with many sharp quills. Its quills come off easily and stick into any attacking animal.

Hibernation During winter in a temperate forest, the temperature is cold. Food is hard to find. Some animals survive by going into hibernation. **Hibernate** means "to rest through winter." Hibernating animals use little energy, and they do not eat. Dormice and bears are two animals that hibernate in winter.

Word Study

The word hibernate comes from the Latin word *hibernare,* meaning "to pass the winter."

Look for the insect in this picture. The thornmimic treehopper looks almost the same as the thorn.

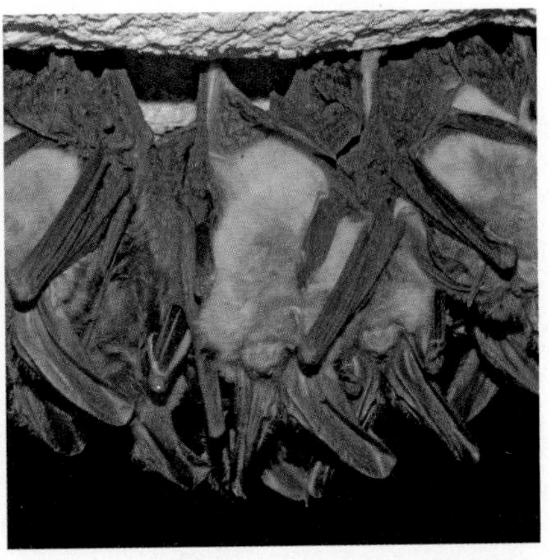

Bats hibernate in sheltered areas in winter.

This skunk has raised its tail and is prepared to spray a predator. Most animals see this and run away.

Ocean Adaptations

Oceans are home to millions of living things. Each living thing has adaptations that help it survive in the salty water.

Algae

Seaweeds look like plants, but they are not. They are plant-like organisms called algae. Like plants, algae make their own food from sunlight. Most algae have structures that are like leaves. Some algae have root-like structures for attaching themselves to the ocean floor. Because they need sunlight, rooted algae can live only in shallow water. Other algae have no roots and drift near the ocean's sunlit surface.

Some algae have balloon-like air bladders. Air bladders help lift the leaf-like structures of rooted algae toward the surface where sunlight is brighter. Air bladders also keep afloat some algae that drift with the currents.

Air bladders help kelp float.

Kelp is a kind of algae. This picture shows a seaweed forest of kelp.

Ocean Animals

Whales and dolphins breathe air. They can hold their breath for a long time as they dive deep looking for food. When they need to breathe, they rise quickly to the surface. They take in air through a hole in the tops of their heads. Fish, on the other hand, have gills for getting oxygen from water.

Many ocean animals have fins. Fins help them swim quickly and control their movement. Ocean animals swim for long distances when they migrate. **Migrate** means "to move from one place to another." Animals migrate to find food, to reproduce, or because the water temperature has changed.

Sunlight cannot reach deep in the ocean. Few organisms live there because the water is very cold. An angler fish that lives in the deep ocean has a light on top of its head. Other animals see the light. They come toward it and become food for the fish.

Did You Know?

Tube worms are adapted to living near hot springs on the ocean floor, far below the surface.

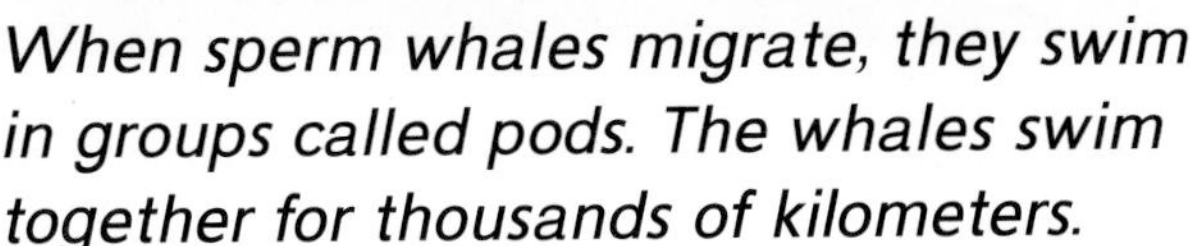

When sperm whales migrate, they swim in groups called pods. The whales swim together for thousands of kilometers.

In the darkness of the deep ocean, the angler fish's lighted "fishing pole" attracts prey.

Wetland Adaptations

In a wetland, the land is often soggy or even under water. At other times it is dry ground. Wetland organisms have adaptations for all of these conditions.

Did You Know?

Many migrating birds stop at wetlands along their flight paths for rest, food, and water.

Wetland Plants

Wetland plants can survive big changes in water level. Mangrove trees live in wetlands alongside rivers or oceans. They are not washed away because their roots are spread out and grip the ground firmly.

Wetland Animals

Wetland animals have adaptations for when the land is covered with water. They can also live when the land is dry. A walking catfish normally lives in a pond. When the pond dries up, the fish moves to another pond by pulling itself over the land using its fins.

This mangrove tree's roots rise up above the water.

This walking catfish has adaptations to breathe air during its short trip over the ground.

Traits

How would you describe a lion? A lion has two ears, four legs, and a tail. It hunts and eats other animals. Its body is covered with fur. Each of these is a trait of lions. A trait is a feature of a living thing. Eye and hair color are examples of traits.

Inherited Traits

An apple tree has roots, a trunk, and leaves because its parents did. Heredity is the passing on of traits from parents to offspring. Features that come from parents are called inherited traits. Most living things inherit traits from both parents. An apple tree's ability to make apples is an inherited trait.

An organism's inherited traits show us what kind of living thing it is. We know a zebra from a monkey by its traits. Those traits are adaptations to help it survive. Zebras inherit legs that help them run quickly. Monkeys inherit arms and legs just right for climbing trees.

Skill Builder

Read a Photograph
Compare the young polar bear and its parent to identify inherited traits.

Design Pics/John Pitcher

Inherited Traits - Individual Traits

You and your friends are all the same kind of living thing. You are all human beings. But you do not all look exactly the same. Some people are taller than others. Individuals have different shades or colors of hair, skin, and eyes. These individual traits are passed from parents to offspring.

Often, parents have different individual traits from each other. Perhaps one parent has blue eyes and the other has brown eyes. Their offspring will inherit some traits from one parent and some from the other. For example, a child might have the mother's hair color and the father's nose shape.

Fact Checker

An offspring can have traits different from both parents. Some inherited traits do not show up in the parents but do appear in the offspring.

Design Pics/Con Tanasiu

Red tulips and yellow tulips can have offspring that are red, yellow, or a mixture of both.

Learned Traits

Some of your traits come from your parents, but others are learned. People and animals can gain new skills over time. These new skills are called learned traits. Learned traits are not passed from parents to offspring.

A sea otter smashes a clamshell with a rock. This learned trait helps the otter get food.

Children are not born knowing a language. They learn to speak from others in their environment. The traits of being able to ride a bicycle, play piano, or throw a ball are learned. These abilities usually improve with practice.

Organisms can be changed by their environments. If a plant's green leaves get too much water, they may turn yellow. Writing with a pencil may cause calluses to develop on your fingers. When a rabbit has plenty of food, it may grow fat. When food becomes scarce, the rabbit may become skinny.

Constant, strong winds changed the shape of this tree.

(t) David Gomez/Getty Images; (b) Ingram Publishing/SuperStock

Inherited Behaviors

A **behavior** is a set of actions. An *inherited behavior* is a set of actions that parents pass on to their offspring. Robins build nests in which they lay their eggs and raise their young. All robins build their nests in the same way. Other birds also build nests, but their nests look different. Each type of bird inherits the information for building its nest. They know how to build their nests without being taught how.

An instinct is a way of acting that an animal does not have to learn. Most animals have an instinct for getting food. A spider web is very complicated, yet a spider is born knowing how to spin one. The spider just does it because building a web is an instinct. Other instincts help animals stay safe and care for young.

This spider has an inherited behavior of building a web for catching food.

Learned Behaviors

A *learned behavior* occurs when experience leads to a change in how an animal acts. Most animals can change how they respond to their environment. The ability to learn new behaviors helps animals survive. A sparrow eats birdseed in a backyard bird feeder. Later, the sparrow comes back to eat there again.

One way animals learn a new behavior is by watching and imitating others. A young prairie dog observes adult prairie dogs being watchful for predators. The young prairie dog learns to listen for the warning call from an adult that signals danger. It imitates the adults and jumps safely into its burrow.

Animals learn new behaviors through life experiences. A young coyote does not know how to hunt for food. It goes on hunts with adult coyotes and learns how to act when hunting. It also learns how to be a member of a coyote pack.

Baby ducks watch and imitate their mother. They learn what to eat and how to be safe.

Humans and the Environment

Natural Resources

A *natural resource* is something in nature used by living things. People use many resources from nature.

Renewable and Nonrenewable Resources

Some natural resources, called renewable resources, will never run out. A *renewable resource* is a resource that can be replaced or used again and again. Plants, animals, water, and air can be replaced. New plants are grown. New animals are born or hatched. Rain and snow bring more water. Plants put oxygen back into the air. Plants, animals, water, and air are renewable resources.

Trees are a renewable resource. Areas of a forest that are cut down can be replanted.

Other natural resources are nonrenewable. A *nonrenewable resource* is a resource that cannot be replaced or reused easily. Metals are nonrenewable. Once they are mined, no more will form. Soil is also a nonrenewable resource. It takes at least 100 years for 2.5 centimeters (1 inch) of soil to form where conditions are good. In places with little rainfall and few plants, it can take much longer for soil to form.

Soil is a nonrenewable resource. Farmers carefully plow their fields to keep soil from being washed away.

Energy Resources

Most of the energy people use comes from fossil fuels. Coal, oil, and natural gas are *fossil fuels.* These fuels formed millions of years ago from the remains of plants and animals. Fossil fuels are nonrenewable. Once they are gone, they will be gone forever.

Solar panels like these gather the Sun's energy and change it into electricity.

People can choose to use renewable sources of energy. The Sun provides Earth with a renewable source of energy every day. A tool called a solar cell can change the energy from sunlight into electricity. Windmills can harness the energy in moving air. Flowing water is also a source of energy. Ocean tides carry usable energy. Dams channel the energy of flowing water in rivers. In some places, people get energy from the heat inside Earth.

This drilling platform drills for oil. Oil is an energy resource and is used to make products.

Heat from inside Earth can be used as a source of power.

Natural Resources

Water

Water is a renewable resource. About 70 percent of Earth's surface is covered with water. Most of this water is held in oceans. This water contains salt and other minerals. People cannot drink this salty water. Only about three percent of Earth's water is *fresh water*. Fresh water can be found naturally in streams and rivers, most lakes, glaciers and icebergs, and in the ground.

Fresh water also falls from clouds as rain or snow. This water replaces the water people take from rivers and lakes for their use. We also get fresh water when snow melts high in the mountains and runs down in streams. Like air, moving water can be turned into electricity.

Humans need to drink fresh water in order to survive. Farmers use water to help crops grow. Some factories use water when they make things. Humans also fish, play, and swim in water.

Most of Earth's water is salt water found in its oceans.

Freshwater Sources

Most fresh water on Earth is locked in glaciers and ice caps. Only a small portion flows in rivers or is held in lakes. Fresh water is a limited resource. People have found many ways to get the water they need.

Rivers and Lakes Many cities draw water from rivers and lakes. Before this water is sent to homes, it is treated to remove bacteria and other harmful things. In some places, people build a dam across a river. The water behind the dam forms a lake. The lake stores water for people to use.

Aqueducts and Wells In some places pipes or ditches, called aqueducts, carry water to where it is needed. Some of Earth's fresh water is found underground. To reach this water, people dig wells. Water is then pumped up from the wells when it is needed.

Desalination Most of Earth's water is salty. Desalination, the process of removing salt from ocean water, produces fresh water that people can use.

What Could I Be?

Desalination Engineer

All over the world, people are looking for ways to increase freshwater resources. Desalination engineers work to improve methods of making seawater drinkable. Learn more about a career as a desalination engineer in the Careers Section.

Thinkstock/Stockbyte/Getty Images

Natural Resources - Air

Air is another renewable resource. Air surrounds our planet. Moving air brings clouds from which rain or snow may fall.

Air is a resource people and animals need in order to breathe. You could live without food or water for several days, but you would live only minutes without air. Plants use carbon dioxide gas from the air as they make food. During photosynthesis, plants make oxygen and release it back into the air. Oxygen is the part of the air your body uses.

Cars and factories can make the air dirty. Dirty air can make eyes itchy. It can cause breathing problems. It can harm plants and buildings. Plants, wind, and rain help clean the air to make it safe to breathe again.

Did You Know?

When resting, an adult human breathes about 12–20 times per minute.

The 35-meter (116-foot) blades of these modern windmills spin, changing the energy of moving air into electricity.

©Ocean/Corbis

Environmental Changes

Environments change all the time. Some changes are natural. Floods, wildfires, and tornados are natural causes of change. Living things also change environments as they meet their needs. A spider spins a web to catch insects for food. A bird builds a nest for shelter. A plant takes in water from the soil. These actions change an environment in small ways.

Other living things make bigger changes. Bacteria and fungi break down dead organisms, which adds nutrients back to the soil. Plants use these nutrients to grow.

People change environments more than other organisms do. Some changes, such as planting trees, are helpful. Other changes, such as pollution, can be harmful.

Prairie dogs burrow in soil, making tunnels and chambers where they raise their young. The largest recorded prairie dog town covered 65,000 square kilometers (25,000 square miles)!

Environmental Changes - People and Land

People use land in many ways. People build on land. They use resources from the land to make buildings. Metals, gemstones, and coal are mined from the land. Land is also used for farming.

Whenever people use land, they change it in some way. Building shops and homes changes the land. In some areas, people drain wetlands and build over them. Wetlands help filter water, so pollution increases when wetlands are cleared. People cut down trees to make wood products. Removing forests causes living things to lose their homes. Soil can wash away without tree roots to hold it in place.

Pollution can also change the land. **Pollution** is the release of harmful materials into the environment. Pollution is caused by nature and humans.

What We Throw Away

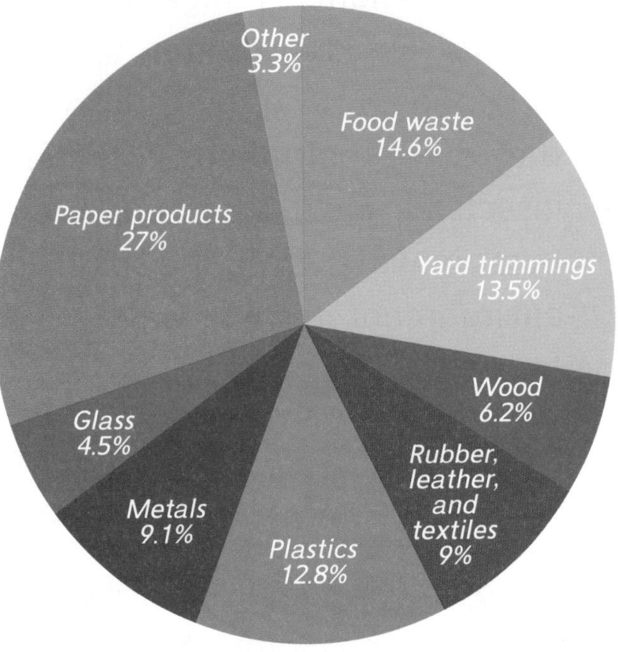

About 110 million metric tons (121 million tons) of waste ends up in landfills every year in the United States.

Ingram Publishing/SuperStock

Air Pollution

Pollution affects air. Volcanoes and wildfires pollute air with dust, gases, and ash. When people burn fossil fuels, they pollute the air. Air pollution of any kind can make it hard for people to breathe.

Water Pollution

Rain can wash the pollution that is in air and on land into water. This pollution can include eroded soil and trash. People who dump trash, oil, and other pollutants into the water also cause water pollution. Polluted water can kill plants and animals. It can make people sick.

Even chemicals that help people can pollute the water. Fertilizers and pesticides can be washed into nearby waters during a storm. People use these chemicals to help plants grow. If these chemicals soak into the ground, they can pollute groundwater. If they wash into rivers and lakes, they can pollute those water resources too.

Sometimes chemicals enter the water as a result of an accident. Oil spills from ships can pollute water and beaches. Many fish, birds, and mammals are harmed in this way.

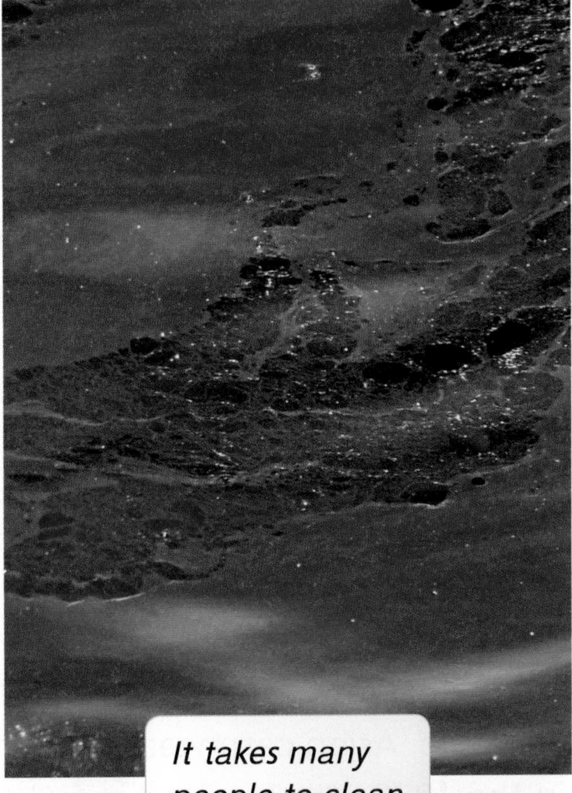

It takes many people to clean up an oil spill.

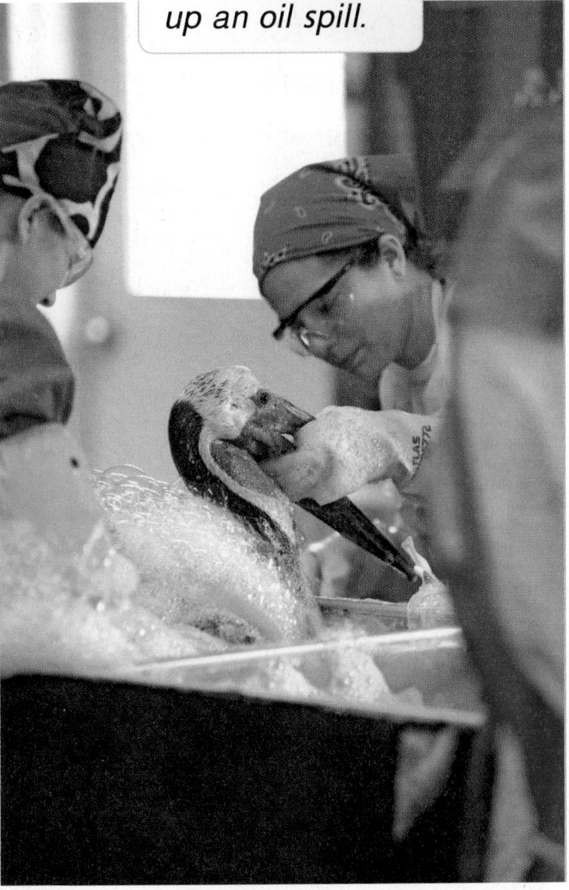

(t) US Coast Guard Photo by Petty Officer 1st Class Tasha Tully; (b) Aaron Roeth Photography

Environmental Changes - People and Living Things

Competition is the struggle among living things for resources. Sometimes people bring new species into an environment. These species often harm the environment. If they have no natural enemies in the new environment, the new species might increase in number. If this happens, the new organisms can use up most of the limited resources in an environment. They crowd out the native organisms.

Most of these species arrive by accident. They are carried on ships, boats, trailers, packing materials, and clothing as people move from place to place. Other species have been brought here on purpose. For example, Asian carp were brought to the United States in the 1970s to be raised in fish farms. Some of these fish escaped. Over time they have spread north. These fish eat the food native fish once ate.

Kudzu plants were brought to the United States from Japan. Kudzu grows rapidly. It uses the water and nutrients that other plants need in order to survive.

Conserving Resources

People change environments in ways that hurt organisms, including themselves. However, people can also help environments. When people *conserve* resources, they use resources wisely instead of wastefully. Below are some ways in which people can conserve.

Saving Energy There are many ways humans can reduce their energy usage.

- Wash only full loads of dishes and clothes.
- Air dry dishes and clothes.
- Turn off lights, computers, and other appliances when you leave a room.
- Ride a bike instead of using a car.
- Unplug chargers from the wall.
- Turn down the heat and wear a sweater in winter.

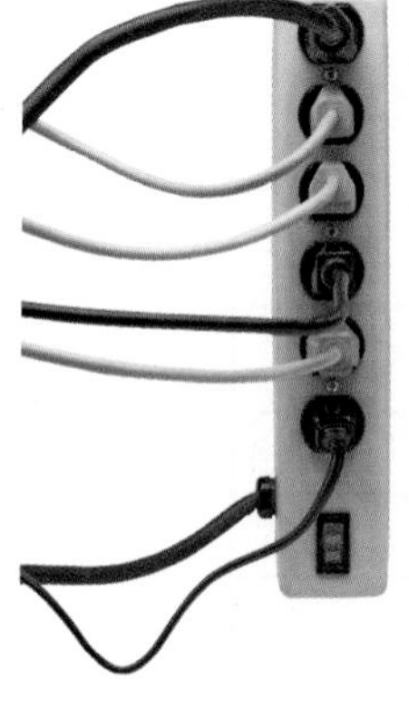

Televisions and other appliances use power even when they are turned off. This standby power *can add up to 5 to 10 percent of home energy use. Flipping off the power strip they are plugged into completely turns them off.*

Word Study

The word *conserve* comes from the Latin *conservare. Servare* means "to keep or guard." When people conserve, they keep natural resources from being used up and *guard* the environment.

What Could I Be?

Energy Auditor

Are you interested in helping families and businesses save money? Energy auditors evaluate homes and businesses for ways to reduce energy use. To learn more about becoming an energy auditor, turn to the Careers Section.

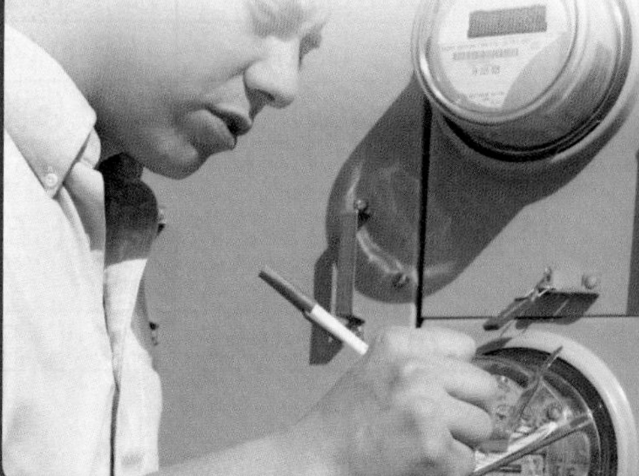

Conserving Resources

Saving Water A family of four in the United States uses about 1,135 liters (300 gallons) of water every day. There are many ways to reduce water use:

- Turn off the water while you brush your teeth.
- Take shorter showers.
- Tell an adult if a faucet is dripping or a toilet is running.
- Run dishwashers or washing machines only when they are full.
- If your family has a garden, use mulch to help keep moisture in the soil. The mulch will reduce the need to water in dry weather.
- When you do water the garden, water in the early morning. This reduces the amount of water that evaporates.

A leaky faucet wastes about 56 L (15 gal) of water a day.

Ways People Use Water

Activity	Normal Daily Use
Showering	95 L (25 gal)
Taking a bath	150 L (40 gal)
Brushing teeth	18 L (5 gal)
Washing hands	8 L (2 gal)
Running a dishwasher	55 L (15 gal)
Washing clothes	220 L (58 gal)

Skill Builder **Read a Table**

This table has two columns. The left column is titled *Activity*. That means everything on the left side is an activity. Everything on the right side tells you how much water that activity uses.

Reduce, Reuse, Recycle

People can also practice conservation by following the three Rs: reduce, reuse, recycle.

When you *reduce*, you use less of a resource. Taking a shorter shower reduces the amount of water used.

When you *reuse*, you use a material again. Use cloth bags when shopping instead of getting new plastic bags. If you do not have a cloth bag, you can reuse old plastic bags.

To *recycle* means to turn old things into new things. Old aluminum cans are melted and made into new cans. Discarded plastic bottles are made into fiber for use in t-shirts and carpet. Some is even used to make plastic lumber for decks and other outdoor uses.

When people walk or ride a bike instead of driving a car, they reduce the amount of gasoline they use.

This home was built from reused glass bottles.

Recycling is one way that people can help protect the environment.

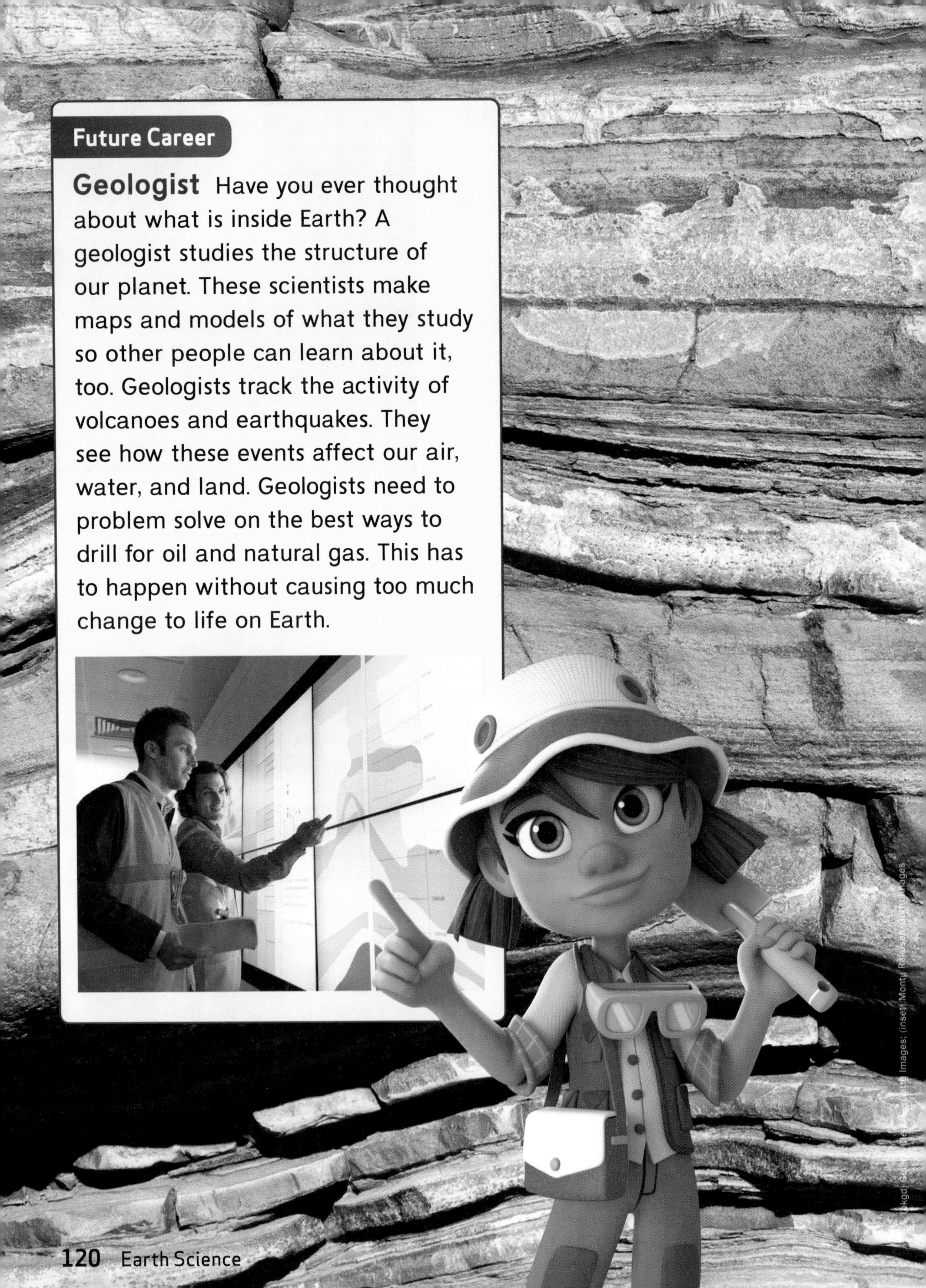

Future Career

Geologist Have you ever thought about what is inside Earth? A geologist studies the structure of our planet. These scientists make maps and models of what they study so other people can learn about it, too. Geologists track the activity of volcanoes and earthquakes. They see how these events affect our air, water, and land. Geologists need to problem solve on the best ways to drill for oil and natural gas. This has to happen without causing too much change to life on Earth.

Earth and Space Science

Earth's Structure

Earth's Layers

If you could look inside Earth, you would see several layers. We live on the outermost layer of Earth, which is called the *crust.* All of Earth's land and the ocean floor make up the crust. The crust is the thinnest of Earth's three layers.

Below the crust is a very thick layer called the *mantle.* The part of the mantle closest to the crust is solid rock. The deepest parts of the mantle are nearly melted rock. In the mantle, the rock is soft and flows a lot like putty.

Earth's *core* is the center of the planet. It is the hottest layer. There are two parts to Earth's core. The center is called the *inner core*. It has the properties of a solid. The *outer core* behaves like a liquid.

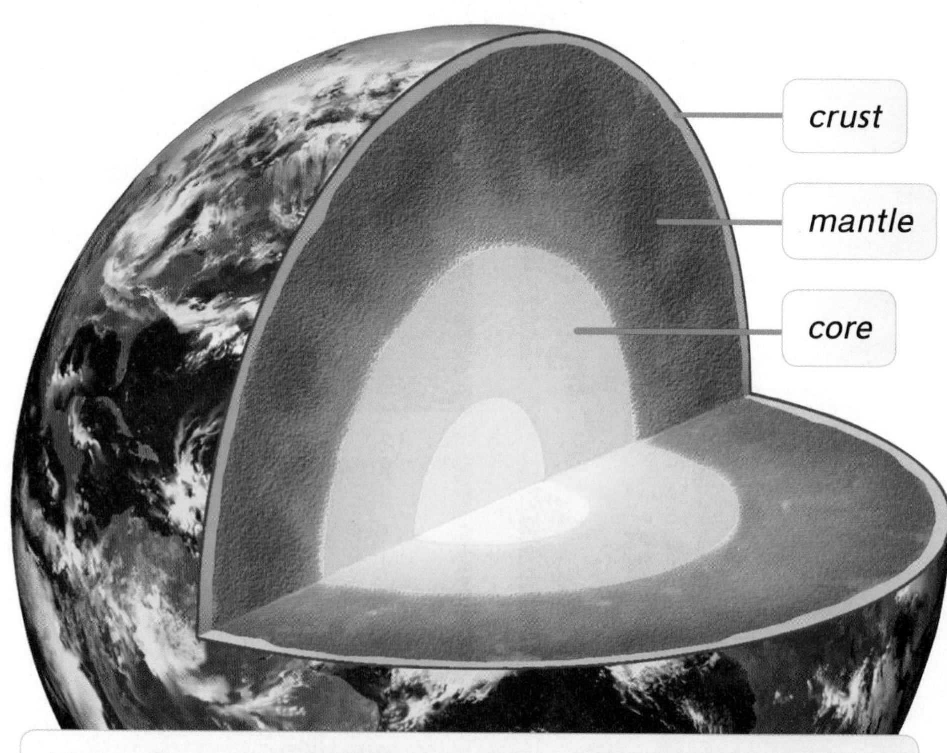

The crust, mantle, and core make up Earth's three layers.

Earth's Landforms

The land around you may have many shapes. It may be flat or rocky. It may slope gently or have jagged cliffs. It may be surrounded by water on some or all sides. Earth's land has many features, called *landforms*.

Plains A plain is a wide, flat area on Earth's surface. Plains that get little rain are covered in short grasses. Plains that get more rain may have trees. Rivers wind through many plains.

Plateaus A plateau is land with steep sides and a flat top. Streams and rivers can cut deep valleys in plateaus. These valleys can become canyons.

Mountains Earth's highest landforms are mountains. High landforms that are smaller than mountains are hills. A group of mountains is called a mountain range.

The middle section of the United States is called the Great Plains. Farmers grow wheat, corn, and oats in the rich soil.

The Grand Canyon has been cut down through the Colorado Plateau.

Mount Everest is the world's highest mountain. It is 8,850 meters (29,035 feet) high.

Earth's Landforms

Some types of maps show where Earth's landforms are located. This map shows the shape of Earth's surface. Each landform took many years to form. Slow processes of Earth build mountains, plains, and plateaus.

Features of Earth

1 A *mountain* is the tallest landform. It often has steep sides and a pointed top.

2 A *valley* is the low land between hills or mountains.

3 A *canyon* is a deep valley with steep sides. Rivers often flow through them.

4 A *plain* is land that is wide and flat.

5 A *lake* is water that is surrounded by land.

6 A *river* is a long body of water moving across land.

7 A *plateau* is land with steep sides and a flat top. It is higher than the land around it.

8 A *coast* is land that borders the ocean.

9 A *peninsula* is land surrounded by water on three sides.

10 An *island* is land with water all around it.

What Could I Be?

Geologist

Want to become an expert in Earth's landforms? Geologists measure the height of different landforms. They study the materials that make up the landforms. They study how landforms change over time. Learn more about a career as a geologist in the Careers section.

M.C. Larsen/USGS

Minerals

A *mineral* is a solid, nonliving substance found in nature. Table salt, gold, and iron are minerals. Minerals are the building blocks of rocks. They are found underground and in soil. They are found dissolved in water and on the ocean floor.

There are more than 3,000 different kinds of minerals. When scientists try to identify a mineral, they look at its properties. Four of these properties are color, streak, luster, and hardness.

Color Color is the first thing people notice about a mineral. Some minerals are only one color. Gold is always a yellow color. Turquoise is always a shade of blue-green. Color alone cannot identify every mineral, however. Quartz comes in many colors.

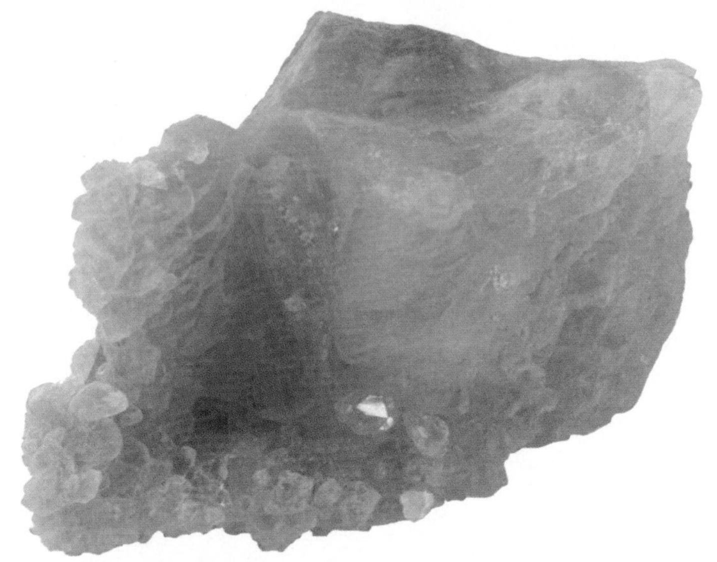

Color isn't always the best way to identify a mineral. All of these minerals are quartz.

Streak Another property of a mineral is streak. Streak is the color of the powder that a mineral leaves behind when it is scratched across a rough surface. Two minerals may look alike until their streaks are tested. The color of a mineral's streak can help identify it.

Both of these mineral samples are hematite. Even though they look different, they both leave a red streak.

Luster Some minerals are shiny, and others are dull. Luster describes the way light bounces off a mineral. If two samples look alike but have a different luster, they are probably two different minerals.

Pyrite is shiny. It has a metallic luster.

A diamond is so hard that it can be used to cut objects.

Hardness The hardness of a mineral describes how easily it can be scratched. Minerals, such as talc and gypsum, are soft. They can be scratched with a fingernail. Quartz is much harder. Not even a steel file can scratch quartz.

Rocks

A *rock* is a nonliving material made from one or more minerals. The mineral pieces that make up rocks are called grains. A rock's texture is how its grains look. Rocks with large grains have a coarse texture. Rocks with very small grains have a fine texture. There are hundreds of different types of rocks. Rocks are classified by how they form.

Igneous Rocks

Igneous rocks form when melted rock cools and hardens. Inside Earth, melted rock called *magma* cools and hardens very slowly. A rock with large mineral grains, such as granite, forms once magma has cooled and hardened. Melted rock that flows onto Earth's surface is called *lava*. Lava cools and hardens quickly. A rock with small mineral grains, such as basalt, forms once lava has cooled and hardened.

Fine-grained rocks form from lava on the surface. If magma in the chambers underground were to harden, it would form coarse-grained rock.

Sedimentary Rocks

Sedimentary rocks form from layers of sediment. *Sediment* is tiny bits of rock, soil, and once-living things. Sandstone, shale, and limestone are examples of sedimentary rocks.

Sedimentary rocks form where sediments are deposited. This deposition often happens at the bottom of a lake or the ocean. Over time, many layers of sediment build up on top of one another. The top layers press on the layers below. Air and water gets squeezed out. Eventually the sediments become cemented together and form rock.

Sedimentary rock forms in layers.

Some sedimentary rocks contain fossils. Fossils are the traces of once-living things.

Word Study

The word *metamorphic* comes from the Greek words *meta*, meaning "change," and *morph*, meaning "form." Metamorphic rocks change form.

Metamorphic Rocks

Metamorphic rocks form when heating and squeezing change rocks. Rock deep inside Earth becomes very hot and squeezed by rocks above them. Heating and squeezing change the mineral makeup of the rock, causing a new rock to form with properties that are different from the original rock.

Over time, heating and squeezing can turn shale into slate.

The Rock Cycle

Over time, one type of rock can become any other type. The process of rocks changing from one type of rock to another is called the *rock cycle*. The diagram below shows this cycle.

Rock materials are moving through the rock cycle all the time. Wind and water break down all types of rock into sediments. The sediments get deposited somewhere else, and eventually form sedimentary rock. Any rock can melt and cool to form igneous rock. Any rock can be heated and squeezed to become metamorphic rock. In time, wind and water break those rocks apart, and the cycle continues.

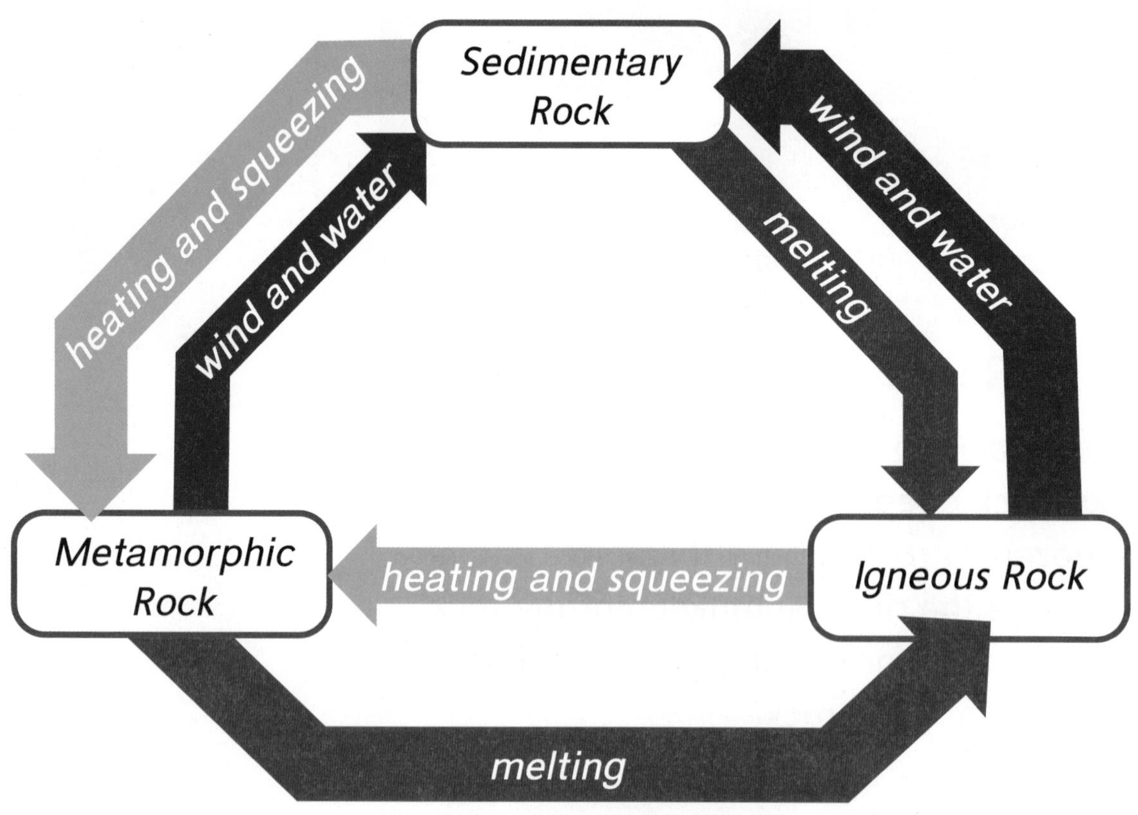

Skill Builder **Read a Diagram**

Follow the arrows to see how each kind of rock can change.

Using Minerals and Rocks

You use minerals every day. Toothpaste is made from the mineral fluorite. Halite is used to make table salt. Calcium is also a mineral. The calcium in milk helps keep bones strong. The glass that holds the milk is made with quartz.

Each mineral has uses because of its properties. Hard minerals such as diamonds are used for jewelry. Because they are so hard, they are also used in large drills. Graphite makes up the "lead" of pencils. This mineral is soft and easily makes a line on paper.

Rocks are also used in many ways. Rocks are used for building roads, houses, and statues. Limestone is used to make cement. Coal is burned for heat.

The aluminum that is used for baseball bats comes from a mineral called bauxite.

Marble is a hard rock that lasts well in all kinds of weather.

Water on Earth

Salt Water

Earth has been called "the water planet." About three fourths, or 75 percent, of Earth's surface is covered with water.

Most of Earth's water is salt water found in oceans. Only a small amount of water on Earth is fresh water. Some of the salt in ocean water comes from rock on the ocean floor. Other salt is washed off the land and runs into the oceans. The salt mixes with the water. You cannot drink or water plants with ocean water. People cannot use this water to make products in factories.

About 75 percent of Earth's surface is covered in water. About 25 percent is land.

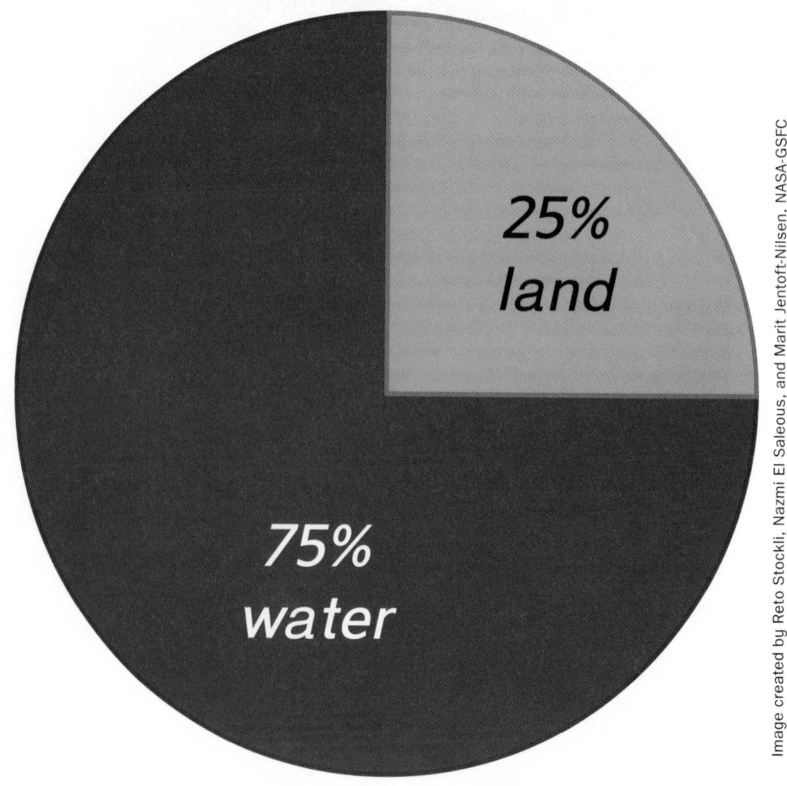

Image created by Reto Stockli, Nazmi El Saleous, and Marit Jentoft-Nilsen, NASA-GSFC

Fresh Water

The water people use is called fresh water. *Fresh water* has very little salt in it. Fresh water is found in many places.

Most fresh water is frozen as ice near the North and South Poles. Glaciers and icebergs are also made of fresh water. *Glaciers* are large, slow-moving masses of ice. They form in places where a lot of snow falls during winter but little melts during summer. After many years, the snow continues to pile up. The snow slowly turns to ice and begins to move.

Many people get fresh water from lakes, rivers, and streams. Streams run together to form rivers. Rivers can flow into and out of lakes.

Some of Earth's fresh water is stored underground. Water seeps down slowly through the soil. It collects in spaces between underground rocks. This water is called *groundwater*. People dig wells to bring groundwater to the surface.

Almost one fourth of Earth's fresh water supply is frozen in glaciers and icebergs.

Water seeps into the ground. It collects in the spaces between underground rocks.

Soil

Soil is a mixture of small pieces of broken rocks, minerals, and other things. It has bits of decayed plants and animals, called *humus*. The humus in soil adds nutrients for plants. Water, air, and living things are also found in soil.

When it rains, water soaks into soil. The humus in soil soaks up rainwater and helps keep soil moist. Plants take in this water through their roots. Roots also help hold soil in place. They keep soil from being washed or blown away.

Some animals live in soil. Worms, ants, and moles move through soil and break it up by separating pieces of rock and humus. They help air and water get into the soil. When the animals die, their bodies decay. They become part of the humus that gives nutrients to the soil.

Earthworms and ants are just a few of the living things found in soil.

How Soil Forms

Soil takes a long time to form—up to 1,000 years for just a few centimeters!

The making of soil starts with weathering, which causes rocks to be broken into small pieces. The tiny bits of rock build up as layers. The lowest layers are made of more solid rock, called *bedrock.* Above the bedrock is a layer called subsoil. The *subsoil* is made of smaller pieces of rock with soil and humus mixed in. *Topsoil* is the top layer that plants grow in. This layer has even smaller rocks and much more humus than the layers beneath it. The movement of animals and plant roots in the soil mixes the topsoil. Over time, parts of the soil get broken down into even smaller pieces.

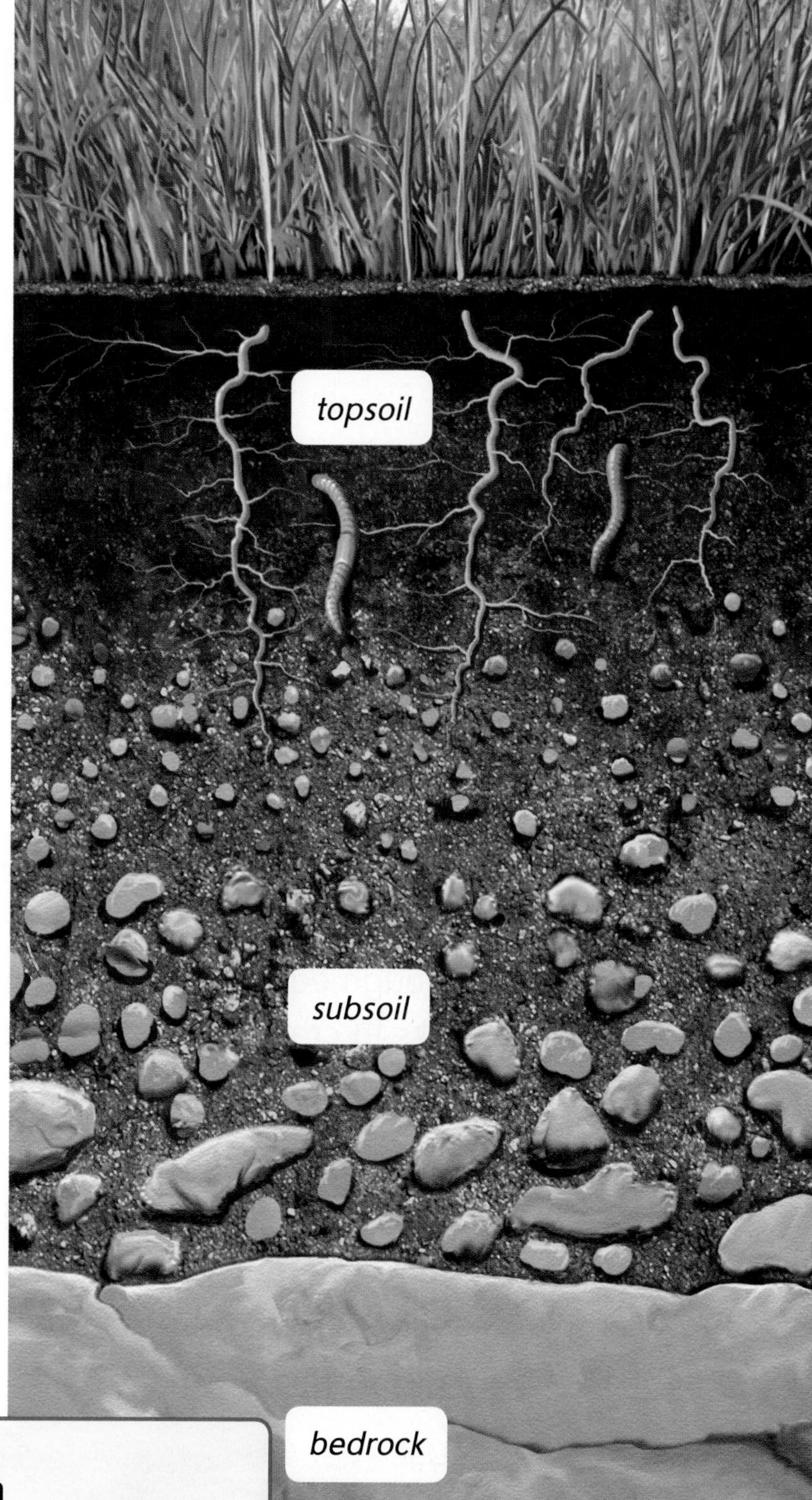

Soil has more rocks in the lower layers, and more humus in the top layers.

Skill Builder **Read a Diagram**

Look at how the color of the soil changes from top to bottom. The darker the layer, the more humus it contains.

Earth's Changing Surface

Weathering

Rain, wind, temperature changes, and running water can slowly change Earth's surface over time. Large rocks break down into smaller rocks. These small rocks break down into sand and soil. *Weathering* is any action that breaks rocks into smaller pieces.

Over millions of years, weathering can change the size and shape of rocks. The changes are usually too slow for people to notice. Waves crash against a cliff and slowly break the rock into smaller pieces. Rivers run across rocks and smooth away rough edges. When water freezes and melts, it causes rocks to crack and break apart. Wind picks up small rocks and scrapes them against other rocks.

Living things can cause weathering too. Plant roots slowly break apart rocks. Animals dig into the ground and expose rock to rain and wind. Weathering caused by water, wind, temperature changes, and living things is called *physical weathering.*

Waves crash against the cliff and slowly break the rock apart.

Physical Weathering

Freezing and Thawing The rock formations shown here are called hoodoos. They have been weathered by freezing and thawing. Water enters small cracks in the rock and freezes. It then expands, or takes up more space. The cracks get wider. Then the ice melts and becomes liquid again. Repeated freezing and thawing breaks the rock apart.

Wind In deserts, wind plays a part in weathering. This rock shape called a ventifact was formed when winds picked up sand from the desert and blasted it against the rock. The sand scraped away parts of the rock and left this irregular shape.

Plants You may have seen cracks in a sidewalk near a tree. Tree roots can crack rock just as they crack a sidewalk. Tree roots and other plant roots can grow in the cracks of rocks. As the roots grow, they force the cracks to widen. The tree in this picture continues to break the large rock apart.

Weathering - Mountain Weathering

Weathering breaks down even the tallest mountains. The same forces that break rock apart also cause mountains to weather. Rain, running water, wind, temperature, and freezing and thawing all play a part. Waterfalls flow down mountains and slowly carve and smooth the rock. The Sun heats and cracks rock.

Seeds blow in the wind and are caught in the rocks on mountains. As the plants grow, their roots crack the rock. Rainwater falls into the cracks caused by the roots. When the water freezes, it expands. More cracks are formed. Over time, pieces of rock fall off and are worn down.

Mountains weather by small amounts each year. As they are weathered, they get smaller and their surfaces change.

Did You Know?

The Great Smoky Mountains in the southeastern United States were once as tall as the Rockies. Weathering has made them smaller and their peaks more round.

The Great Smoky Mountains

The Rocky Mountains

The Great Smoky Mountains formed between 200 and 300 million years ago. Their peaks have worn down over time. The Rocky Mountains are younger than the Smokies. Their peaks are taller and sharper.

Chemical Weathering

There is also another type of weathering. *Chemical weathering* changes the minerals that make up rocks. Water and air can change iron in rocks into rust. Gases in the air from cars and factories can combine with rainwater to form acid rain. This rain eats away at the surface of rocks. Acids given off by some living things can also cause the breakdown of rocks.

Look closely at the noses on these statues at the Acropolis in Greece. They did not fall off. They were dissolved by chemical weathering.

Glow Images

How Weathering Changes Rock	
Causes of Weathering	**Examples**
Rain and running water	Waves break apart rock along the shore.
Wind	Wind blows sand and small rocks against larger rocks.
Freezing and thawing	Cracks widen as water freezes and expands.
Temperature changes	Heating and cooling cause rocks to crack.
Plants and animals	Roots break apart rocks.
Chemicals	Acid rain eats away the surface of rocks.

Erosion

Sometimes rock pieces are moved from one place to another. *Erosion* is the movement of weathered rock. Wind, water, gravity, living things, and ice cause erosion.

> **Did You Know?**
>
> The smooth rocks that are often seen on the banks of a river are called river rock. They have been weathered smooth by smaller bits of rock carried by flowing water.

Erosion by Water and Wind

Moving water is a major cause of erosion. Rainwater flows over the land. It can carry away soil from farm fields. Waves crash against the shore and pick up sand and small rocks. The waves carry the pieces to other places. Rivers pick up bits of rock and carry them from place to place. Eventually, the rocks and sand, or sediment, are dropped in new places. *Deposition* is the dropping of sediment in a new location.

Erosion by wind is common in dry areas. Few tall plants grow in very dry areas, so there is little to slow the wind. Wind picks up and carries away the dry soil. When wind speed slows, the soil is deposited in a new place.

Water slows as it moves into the ocean. Sediment can be deposited to form a delta.

NASA-JSC

Erosion by Gravity and Living Things

Gravity can cause erosion. Gravity pulls rocks and soil downhill. This material moves slowly on gentle slopes. Weathered material can move very quickly on steep slopes. A mudflow is the quick movement of very wet soil. A rockslide is the quick movement of rocks down a slope.

Living things can also cause erosion. Prairie dogs tunnel through soil. Worms mix and move soil. Ants and bees move soil to make underground nests. Erosion continues as water moves through the tunnels and spaces made by living things.

These rocks were pulled down by gravity.

These railroad tracks did not start out crooked. Gravity causes the soil on the hillside to slowly move downhill. This type of slow erosion is called creep.

(t) Design Pics/Jack Goldfarb; (b) W.W. Atwood/USGS

Erosion - Erosion by Ice

Frozen water can also cause erosion. A glacier is a mass of ice that moves slowly across the land. As it moves, a glacier picks up and carries away rocks of all sizes. The ice at the bottom of a glacier freezes on the rocks. As the glacier moves, it tears rocks out of the ground. A glacier can move rocks as large as a house! As the glacier melts, it deposits the rocks in a new place.

There have been times when Earth was much colder than it is today. There were many more glaciers than there are now. Many of those glaciers have melted. Valleys and rocky structures are left where glaciers eroded the land.

The dark lines in this glacier are rocks and soil. The glacier picked up this material as it moved. The glacier deposited the materials in a new place as it melted.

Erosion by People

Erosion is a part of nature, but people also do things that cause erosion. Some changes are very small, like digging a hole to plant a tree. Other changes are much larger.

People cut down forests for wood or clear land for farming. If the trees are not replaced, erosion can easily wash away the soil. Mining and other digging into the surface of the land also cause erosion. Dams and structures built on rivers can change the normal process of erosion. In other places, ponds and swamps are drained. The dry soil left behind can blow away.

Some types of mining clear plants from the surface of land. Clearing plants can lead to erosion.

What Could I Be?

Trail Designer

Interested in designing recreational trails that don't cause erosion? A career in trail design might be for you! Find out more about this career in the Careers section.

Volcanoes

A *volcano* is a mountain that builds up around an opening in Earth's crust. Magma is melted rock from parts of the mantle and crust. This melted rock rises up through the opening. When magma flows onto land, it is called lava. Lava can flow slowly over land or bubble up around a volcano's opening.

Some volcanoes explode, or erupt, with more rocks and ash than lava. When a volcano erupts quickly and with a lot of energy, parts of the volcano may be blown apart. The side of a volcano may be blasted away in an instant.

Volcano eruptions change the land in other ways, too. After lava flows along Earth's surface, it cools and hardens. Over time, the hardened rock builds up around the volcano to form a mountain. A volcanic mountain can take many years to form. If a volcano erupts often, however, a volcanic mountain can form in just a few years.

How a Volcano Grows

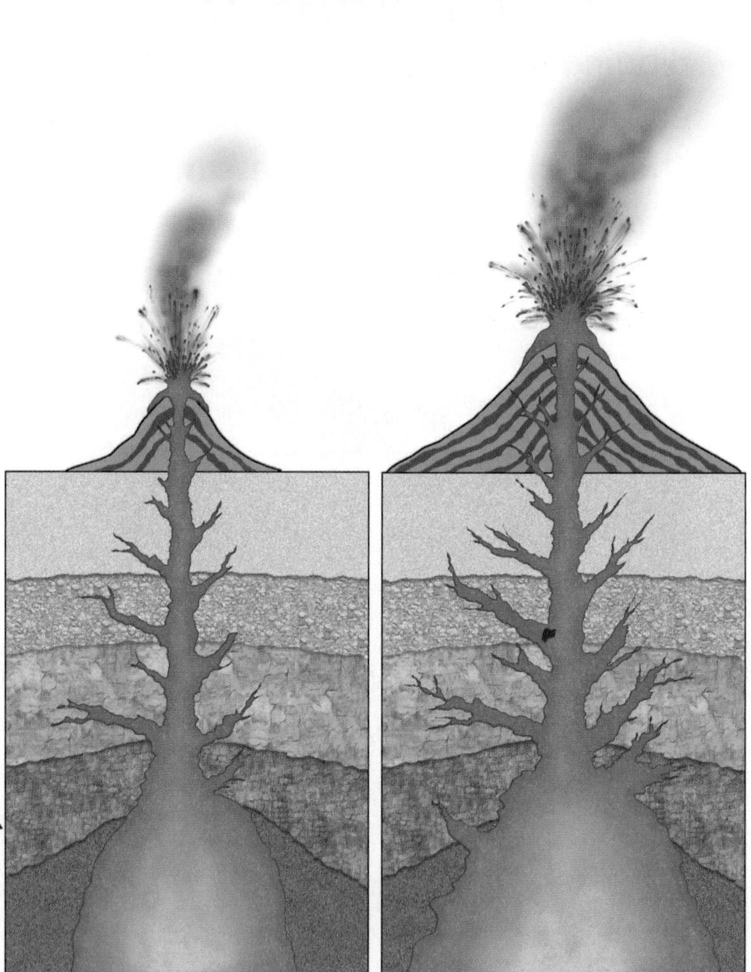

Skill Builder **Read a Diagram**

Compare the two parts of the diagram. Notice how the volcano has changed.

Between 50 and 70 volcanoes erupt on land each year. Volcanoes that erupt often are called *active volcanoes.* Mount St. Helens in Washington is an example of an active volcano. Some volcanoes are not active. Yet, scientists think they may still erupt. These volcanoes are called *dormant volcanoes.* Mauna Kea, one of the volcanoes that make up the big island of Hawaii, is dormant. Scientists think it last erupted about 4,500 years ago. Volcanoes that have not erupted for at least 10,000 years are called *extinct volcanoes.* It can be hard to know if a volcano is extinct or just dormant.

The Hawaiian Islands formed from undersea volcanoes.

Underwater Volcanoes

There are also many active volcanoes on the ocean floor. Underwater volcanoes send out lava, ash, and rock into the ocean. When the lava hardens into rock, a volcanic mountain may rise up out of the sea. The Hawaiian Islands formed from underwater volcanoes. Only one of these islands includes an active volcano today.

What Could I Be?

Volcanologist

Interested in finding out what makes volcanoes erupt? Do you want to learn warning signs that could help keep people safe? You may be interested in a career as a volcanologist. Learn more about what these volcano scientists do in the Careers section.

Earthquakes

An **earthquake** is the shaking of Earth's surface. Earthquakes are caused by movement in Earth's crust. The crust is made of huge slabs of rock. The slabs can slowly move past each other and can press against each other. They can pull apart too. These movements happen along large cracks called *faults.* When the crust suddenly moves along these faults, the area above the fault can shake.

An earthquake can cause more than just a rumble. It causes vibrations at the place where the two plates hit each other. The vibrations move out in all directions at once. The areas closest to the center of the quake have the strongest vibrations. The areas farther away have weaker vibrations.

Some earthquakes are strong enough to make an opening in the ground. Earthquakes can also cause great damage to buildings and roadways.

Measuring Earthquakes

Scientists estimate that several million earthquakes happen every year. Most of these quakes are too weak to be felt. The instruments that scientists use sense about 50 quakes per day. Scientists use two scales to help measure characteristics of earthquakes. The Richter scale measures the strength of the vibrations sent out at the moment the rock slabs move. The scale uses numbers from 1 to 10. An earthquake of 10 is the strongest. It causes the most damage. There may be only one earthquake a year that rates an 8 or higher on the scale. An instrument called a seismograph detects and records earthquakes based on the Richter scale.

The Mercalli intensity scale measures the strength of shaking at a certain location. It is measured by observing the effects on people, structures, and the environment.

What Could I Be?

Seismologist

Scientists around the world are working to predict earthquakes. They study thousands of small quakes that most people do not even feel. They study the big ones, too! Learn more about what seismologists do in the Careers section.

An earthquake causes a vibration at the point where slabs of Earth's crust move. The vibrations get weaker as they travel away from an earthquake's center.

Floods

Water can slowly change the land, or it can change the land quickly. Sometimes, more rain falls than can soak into the ground. Water quickly fills rivers and streams. If a river cannot hold all the water, the water flows over its banks, causing a flood. A **flood** is water that flows over land that is normally dry. Rapid melting of snow can also cause flooding. Floods can be very strong and dangerous. The heavy flow of water can wash away large objects, such as cars.

Floods can change the shape of the land by eroding soil quickly. Even after the flood is over, the land may stay reshaped. A river's course may be changed. Buildings, bridges, roads, and crops may be washed away. Sediment carried by the floodwater may have been deposited on land.

Floods can cover streets and make it impossible for cars to pass.

USGS photo by Don Becker

Landslides

A *landslide* is the rapid movement of rocks and soil down a hill or mountain. Heavy rocks, trees, and land are pulled down the hill by gravity. Landslides can happen quickly and cause damage to property. Landslides may occur when soil loosens because of heavy rain or melting snow. Some landslides may occur without warning. Others can happen after earthquakes shake loose large areas of rock or soil on hillsides.

Landslides can happen on land or the ocean floor. Landslides on hills and mountains can destroy homes and roads. They can bury large areas. Underwater landslides can form giant waves.

A landslide can destroy homes and cause damage to property.

USGS photo by Jim Bowers

The Water Cycle

The continuous movement of water from Earth's surface into the atmosphere and back again is called the *water cycle*.

The water cycle starts with the Sun's energy. As the Sun warms Earth, water becomes warmer and evaporates. It moves into the air as water vapor. The water vapor cools as it rises. It condenses to form clouds in the sky. The water falls back to Earth. The rain, snow, sleet, or hail that falls from clouds is called **precipitation**.

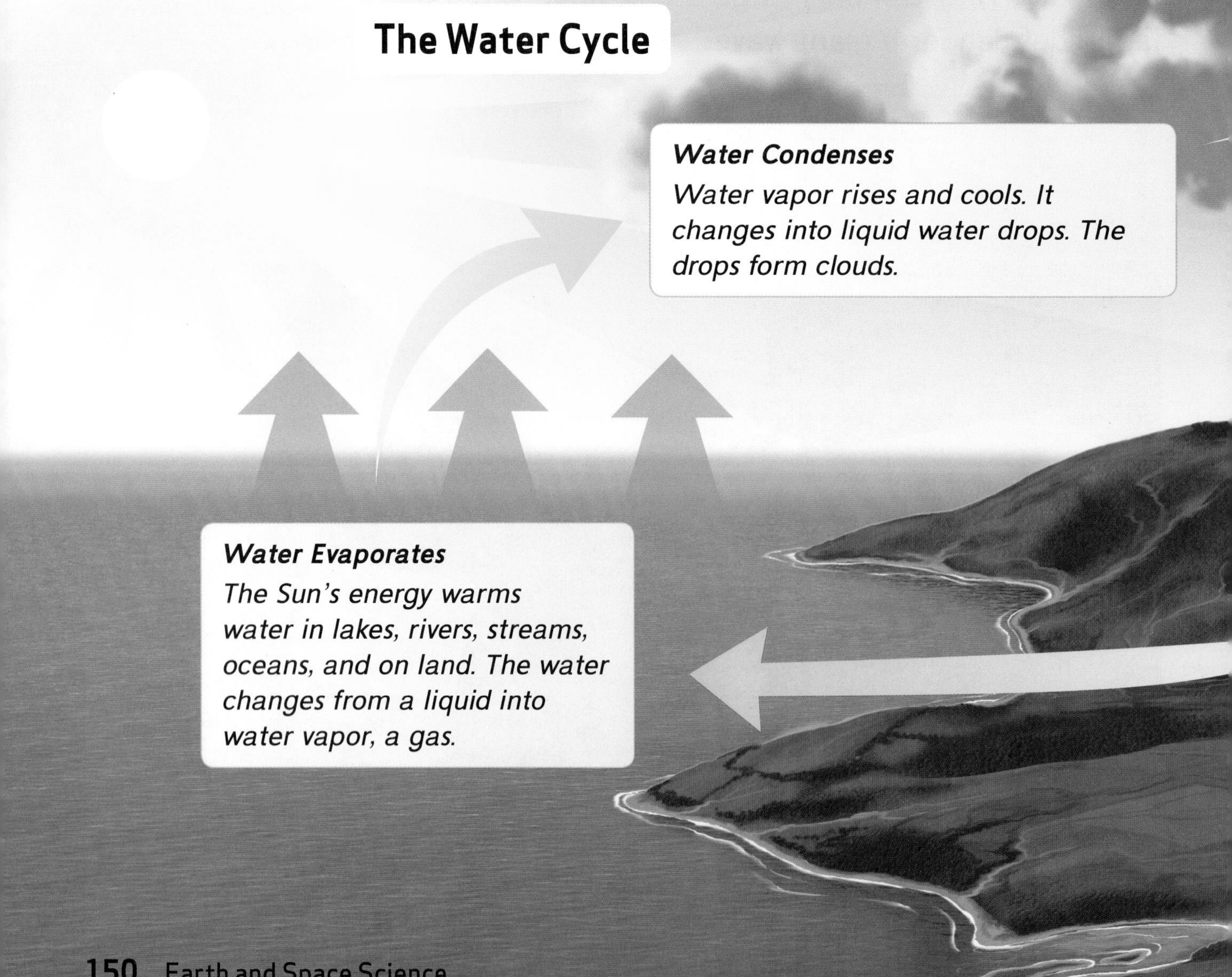

The Water Cycle

Water Condenses
Water vapor rises and cools. It changes into liquid water drops. The drops form clouds.

Water Evaporates
The Sun's energy warms water in lakes, rivers, streams, oceans, and on land. The water changes from a liquid into water vapor, a gas.

Some precipitation falls in oceans. Some falls over land. Water that falls on land may seep into the soil and become groundwater. Some water could flow downhill and may enter a stream, river, or lake. Some of the water will change to water vapor and the process will begin again.

Water Falls

The drops of water in the clouds get bigger and bigger. Finally they fall to the ground as rain, snow, sleet, or hail.

Water Flows

Some precipitation flows over Earth's surface. It collects in lakes, rivers, and oceans. Precipitation can also soak into the ground. Groundwater moves through spaces in underground rock.

Skill Builder **Read a Diagram**

Follow the arrows to understand the processes that make up the water cycle.

Weather and Climate

Weather

Weather is what the air is like outside at a certain time and place. Even though you cannot see air, you can see it move things, such as the leaves of trees. You can feel air as a breeze on your face. Weather changes from day to day. It can also change from hour to hour or even minute to minute.

The air that surrounds Earth is part of the atmosphere. The *atmosphere* is a blanket of gases and tiny bits of dust that surround Earth. The atmosphere has several layers. Weather occurs in the layer closest to Earth.

Mediolmages/Photodisc/Getty Images

Weather includes the clouds in the sky and water in the air. It includes the temperature of the air and how the wind is blowing.

Temperature

When people describe weather, they describe the air and its temperature. *Temperature* is a measure of how hot or cold something is. A *thermometer* is a tool that measures temperature.

The air changes temperature because of energy from the Sun. The Sun heats Earth's land and water. Then the land and water heat the air above them. The Sun's heat is more direct at midday than it is at sunrise or sunset.

Measuring Air Temperature

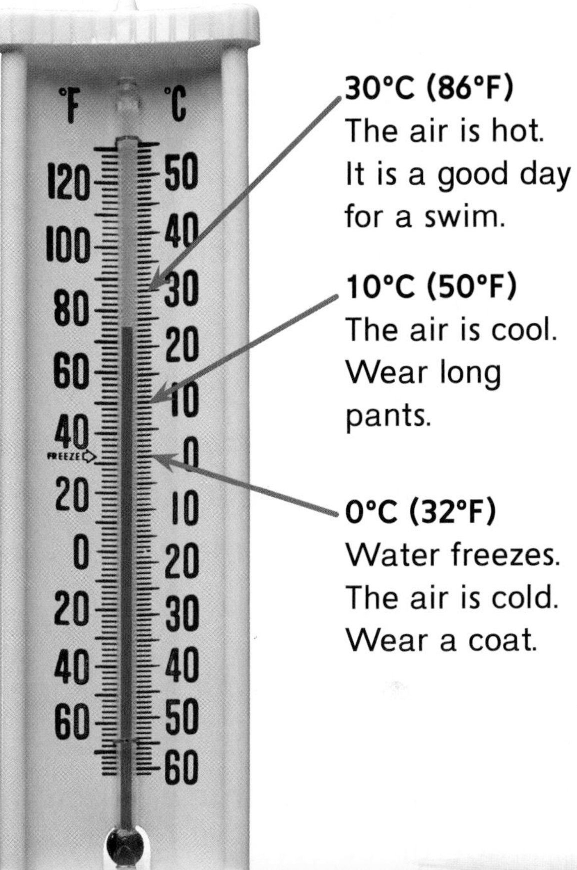

Skill Builder

Read a Diagram

On this thermometer, degrees Fahrenheit are on the left, and degrees Celsius are on the right.

Burke/Triolo/Brand X Pictures/Jupiterimages

Weather - Describing and Measuring Weather

Air temperature is one measurement of weather. Precipitation, wind, and air pressure also describe weather. When one of these factors changes, so does the weather.

Precipitation Precipitation is water that falls to the ground from clouds. Liquid rain is the most common type of precipitation. It falls if the air temperature is warmer than 0°C (32°F).

Sleet, snow, and hail are frozen precipitation. Sleet forms when rain falls through a layer of freezing-cold air. Snow is made of ice crystals. Hail forms when rain freezes and is tossed about in a tall cloud. Layers of ice coat the drops, forming a ball. Eventually, hail gets so heavy it falls to the ground.

The arrow of a weather vane points into the wind.

A rain gauge measures how much precipitation has fallen.

Hail can be as small as a pea or as large as a grapefruit.

Air Pressure **Air pressure** is the force of air pressing down on Earth's surface. Weather reports often describe air pressure. When the air presses down a lot, air pressure is high. When it presses less, air pressure is lower. Low air pressure can mean that weather will be cloudy or rainy. High air pressure often means that skies will be clear. Scientists use a tool called a *barometer* to measure air pressure.

Wind **Wind** is moving air. It is caused by differences in air pressure. Air moves from places where air pressure is higher toward places where air pressure is lower. If these differences are great, the wind will blow hard and move fast. If these differences are small, the wind will blow gently or not at all.

An anemometer measures wind speed.

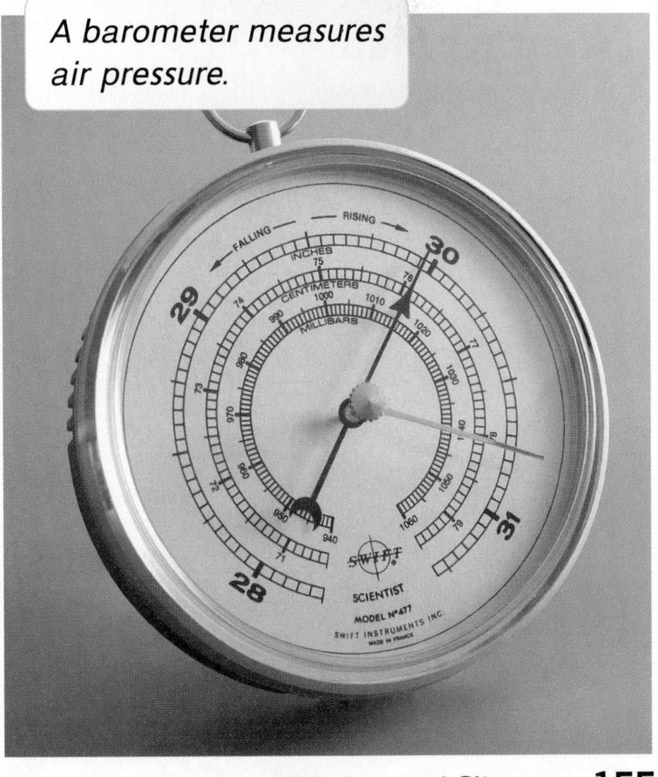

A barometer measures air pressure.

Predicting Weather

Scientists use many tools to find weather patterns and predict how weather will change. They use tools that measure wind speed and direction, precipitation, air pressure, and temperature. A weather balloon is a tool that is launched into the air. It carries devices that collect data about the atmosphere. A satellite is a tool scientists put in space. It travels around Earth and collects weather data over very large areas. A satellite can spot storms in the ocean. Scientists then use other information about the atmosphere to predict where the storm will move next.

A weather map lists the temperatures, precipitation, and other upcoming weather predictions over a large area. It may show the high temperatures for the day. It may show where it will rain and where it will be sunny.

Weather balloons are used to gather data about weather.

Skill Builder

Read a Diagram

Look carefully at the key. The colors show areas where different high temperatures will occur. The symbols show the kind of weather different areas will have.

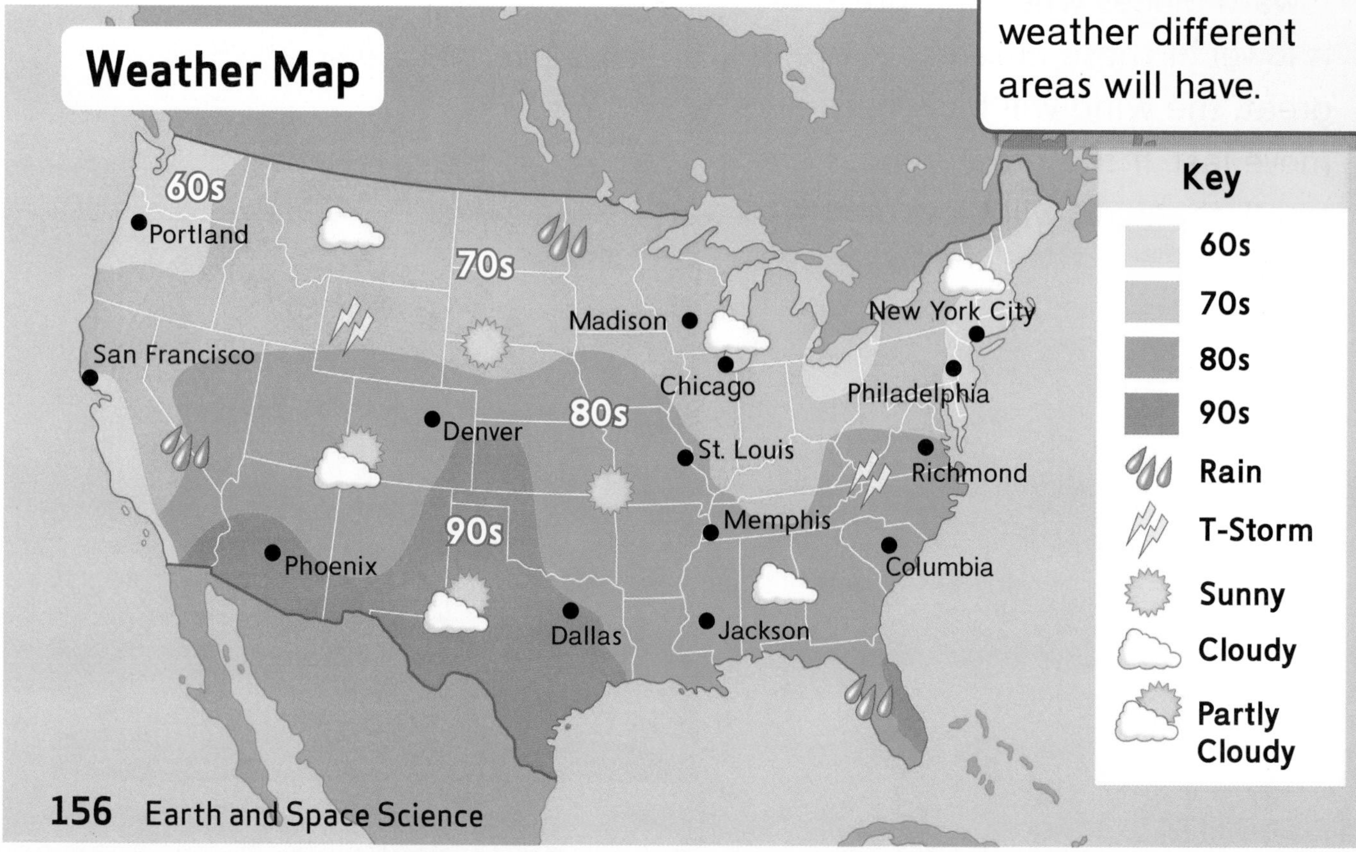

Weather Map

60s
70s
80s
90s
Portland
San Francisco
Madison
Chicago
New York City
Philadelphia
Denver
St. Louis
Richmond
Memphis
Phoenix
Columbia
Dallas
Jackson

Key

60s
70s
80s
90s
Rain
T-Storm
Sunny
Cloudy
Partly Cloudy

Personnel of Quad City National Weather Service Forecast Office

Scientists try to predict what the weather will be to help people plan their lives better. Knowing what the temperature will be helps people decide what to wear. Predictions of rain warn people to wear a raincoat or to take an umbrella. Farmers plan when to plant and how to care for their crops based on weather predictions.

Knowing the weather also helps keep people safe. Airplane pilots study the weather. Weather predictions help them know whether it is safe to take off and then land in another area. Predictions of thunderstorms help people know when to plan indoor activities. Scientists can also predict when severe storms are likely to happen. Predictions of hurricanes can give people time to prepare. They can find shelter. They can leave the area if they have enough time.

What Could I Be?

Meteorologist

Fascinated with the weather? A meteorologist studies weather and predicts upcoming changes in the weather. Turn to the Careers section to find out more about a career in meteorology.

Predictions of severe weather can help people stay safe.

Clouds

A **cloud** is a collection of tiny water droplets or ice crystals in the air. The color and shape of clouds can give information about weather. Clouds can appear white or gray. White clouds are made of ice crystals and reflect light. When clouds get very thick with water droplets, they absorb sunlight and appear dark gray. That's when clouds appear gray. Gray clouds often bring rain. Clouds are moved through the atmosphere by winds. That makes it hard to tell where and when rain may fall.

Cloud shape gives weather watchers a lot of information. A cloud's shape cannot predict weather as well as data from other weather tools. However, each basic cloud type appears under certain conditions in the atmosphere. The three main types of clouds are stratus, cirrus, and cumulus clouds.

Clouds filled with water droplets absorb sunlight, so they look darker than other clouds.

NOAA Photo Library, NOAA Central Library; OAR/ERL/National Severe Storms Laboratory (NSSL)

Stratus clouds *form in low, flat layers. They cover most of the sky, like a blanket covers a bed. Stratus clouds can be white or gray. They often bring rain or snow.*

Cirrus clouds *form high above the ground. They are thin and wispy, and mainly appear white. They show that the weather is fair, but precipitation may come in the next couple of days.*

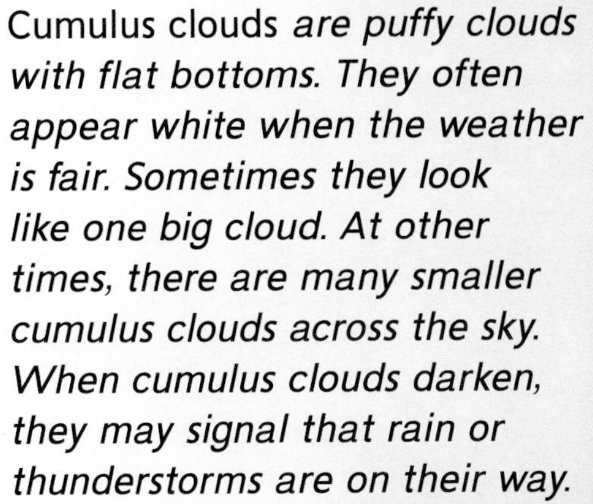

Cumulus clouds *are puffy clouds with flat bottoms. They often appear white when the weather is fair. Sometimes they look like one big cloud. At other times, there are many smaller cumulus clouds across the sky. When cumulus clouds darken, they may signal that rain or thunderstorms are on their way.*

Did You Know?

Earth is not the only planet with clouds. Other planets have clouds made of acids, not water.

How Clouds Form

Clouds are part of the water cycle. As water moves from Earth's surface to the atmosphere, It changes. Water changes from a liquid, to a gas, and back to a liquid again. Sometimes the water freezes into a solid before becoming liquid again.

Clouds contain drops of water that are a result of the process called condensation. *Condensation* is the changing of a gas to a liquid. You may have seen water form on the inside of a window. This liquid water forms when water vapor in the air touches the cooler window. The water vapor cools and turns into liquid water on the glass. In the same way, water vapor in the air outside cools and collects around dust specks. This process forms clouds in the sky.

The drops on this spider web formed as water vapor in the air cooled and condensed.

Ingram Publishing

Like the water in the water cycle, clouds come and go from the sky. The water returns to Earth when it falls as precipitation. The Sun's heat evaporates liquid water into a gas again. The water returns to the sky as water vapor. The process of condensation forming clouds begins again.

All over Earth, different types of clouds are forming. Different parts of Earth are experiencing different types of weather at the same time. There may be puffy cumulus clouds where you live right now. Fifty miles away there may be gray stratus clouds in the sky. Weather is always changing.

Clouds are a reminder that water is always changing form in Earth's atmosphere.

Ivan Rubanov/iStock/Getty Images

Seasons

Seasons are times of the year with different weather patterns. Earth's four seasons are winter, spring, summer, and fall. The north and south halves of Earth have opposite seasons at any given time. The seasons are caused by the way Earth is tilted and moves around the Sun.

Summer has the warmest weather and the most hours of daylight.

During summer, a place has more hours of sunlight during a day. The summer Sun is higher in the sky than at other times of the year. Temperatures are the warmest of the year.

Fall temperatures are cooler than summer temperatures.

In fall, it is daylight for fewer hours and temperatures are cooler. The Sun stays lower in the sky compared to summer. In some areas, leaves change colors and fall from trees. Crops that grew all summer stop growing and are ready for harvest.

On the north half of Earth, summer and fall last from about June to December. Then the season changes again. Weather gets colder as winter arrives.

Skill Builder

Read a Table

Read each season in the first column. Then read the information for each season in the other columns of the chart.

Seasons on the North Half of Earth

Season	Months	Temperatures	Amount of Sunlight
Summer	June through September	highest of year	greatest amount of year
Fall	September through December	getting cooler	decreasing
Winter	December through March	lowest of year	least amount of year
Spring	March through June	getting warmer	increasing

The cycle of the seasons happens in the same order and in the same way every year.

Winter is the coldest season. The Sun's path is lower in the sky. Daylight hours are fewer, and it may be very dark by 5:00 in the evening. Temperatures can be cold in winter. Precipitation may fall as snow in some areas. Winter is too cold for some animals. Many birds move to warmer places. Foxes or rabbits may stay warm in holes under the ground.

In spring, the Sun's path begins to rise higher. Temperatures warm. Trees begin to grow leaves. There are more hours of daylight. Animals that were away during winter begin to return. Many animals have young during spring. Plants and animals have all spring and summer to grow.

In spring, temperatures begin to get warmer.

Seasons are not exactly the same in every place on Earth. Winter in Arizona is warmer than winter in Wisconsin. In both places, however, winter is the coldest season with the fewest hours of daylight for the area.

Winter has the coldest temperatures and fewest hours of daylight.

Climate

Weather changes all the time. It may be rainy one day and sunny the next. But the climate of an area stays the same. **Climate** is the pattern of weather in a certain place over a long period of time. A climate is described by its average temperature and precipitation. One area may have cool, dry summers. Another may have hot, humid summers.

Climates differ based on where an area is located on Earth. Not all areas have four separate seasons. An Arizona desert is hot and dry all year long. Some rain forests in South America are always hot and wet. Many areas have hot and cold temperatures and wet and dry periods throughout the year.

The Sonoran Desert is hot and dry.

Chicago can be hot or cold, dry or wet.

Antarctica is cold and dry.

Rio de Janeiro is hot and wet.

Earth and Climate

Earth is the shape of a *sphere,* or ball. Earth also has an imaginary line called an **axis** through its center. Earth is constantly moving around this axis, like a spinning top. However, Earth's axis is tilted slightly. Earth's axis also points to the same place in the sky all year long. This consistent slant and direction of Earth's axis as it orbits the Sun causes different seasons. This affects climates around the world.

The Sun's rays strike Earth differently at different places because of Earth's shape as well as the consistent axis tilt angle and direction. The Sun's rays strike some places on Earth nearly straight on. These places get the most energy and have warmer climates. The Sun's rays strike other places on Earth at a slant. These places get less energy because the Sun's rays are more spread out. These places have colder climates.

Make Connections

Jump to the Space section to see how Earth spins on its axis and moves around the Sun.

The Sun's Rays

The Sun's rays strike the area of Earth that includes Point A at a steep angle. Point A has a warmer climate than Point B because the Sun's rays are more slanted.

Factors That Affect Climate

Water and Climate

Being near an ocean or other large body of water affects climate. Water absorbs and gives off heat energy more slowly than land. Land heats and cools at a faster rate than water. In summer, ocean water is cooler than nearby land. This tends to keep the air above the land cooler. In winter, ocean water is warmer than nearby land. This tends to keep air above the land warmer than land farther inland.

***Seattle, Washington**, is near the ocean. It has milder temperatures and more rain than places farther inland.*

Large lakes also affect climate. Air blowing across lakes can pick up moisture. The moisture can fall as rain or snow on land areas near the lake.

Height and Climate

How high in the atmosphere a place is affects its climate. Air temperatures get colder as you go higher in the atmosphere. Places in mountains tend to have colder air temperature and climates than low areas.

***Breckenridge, Colorado**, is high in the Colorado Rockies. It has cool temperatures.*

Mountains and Climate

Mountains affect how wet a climate is. One side of a mountain might be wet, while the opposite side might be dry.

Moist air from the ocean moves toward mountains along the coast. The mountains force the air upward. The rising air cools and forms clouds. Rain or snow might fall. This pattern causes places on the ocean side of mountains to have a wet climate.

Air that blows over the mountains is dry. It is dry because the air has lost its moisture on the ocean side. Dry air blows down this side of the mountain. It is common to see deserts on the dry side of a mountain. The mountains block moist air from reaching inland.

What Could I Be?

Climatologist

Interested in worldwide interactions in the atmosphere? Enjoy working with computers? Climatologists collect data and run computer programs. They predict how changes in climate may affect living things, including people. Learn more about a career as a climatologist by turning to the Careers section.

Air cools and loses moisture as it moves up and over a mountain.

***Death Valley, California**, is a dry place. The mountains keep moist air from reaching this area.*

Weather Events

Most of the time, people experience gentle weather, such as a rain or snow shower. There are more severe types of weather as well. Severe weather can be dangerous to property as well as to people, plants, and animals. Weather scientists try to keep track of any severe weather as soon as it starts. They warn people about what to expect in their area. People can stay safe from dangerous weather if they listen to warnings.

Some areas are more likely than others to have certain types of severe weather. Flat lands have different types of severe weather from coastal areas. Areas in colder climates have more snowy weather and ice storms. Different seasons bring different types of severe weather. Keeping an eye out for weather reports can help keep you and your family safe in all kinds of weather events.

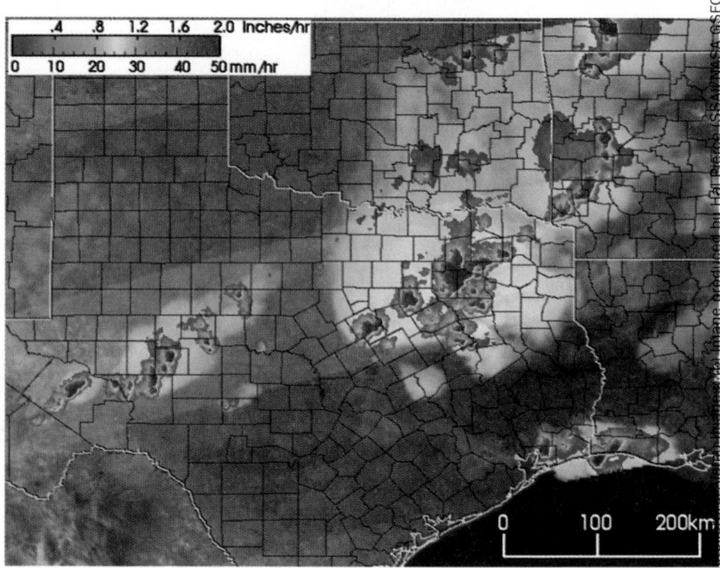

Scientists use tools such as radar to track severe weather. They warn people when to take shelter.

Thunderstorms

A *thunderstorm* is a storm with thunder, lightning, heavy rains, and strong winds. Thunderstorms occur when different types of air come in contact with each other. A mix of warm and cold air, or wet and dry air may cause a thunderstorm to form. This mixing sometimes causes warm, moist air to be lifted. The rising, moist air can result in the formation of tall thunderstorms. The buildup of energy from lifting air may cause lightning. *Lightning* is a static discharge similar to the one you might feel after touching a doorknob on a dry day. *Thunder* is the rumbling sound you hear after lightning quickly heats and expands the air. Thunderstorms can also produce hail, sleet, or snow.

Thunderstorm Safety Stay safe from a thunderstorm by going inside. If you cannot go inside, avoid lightning by staying away from trees and staying out of water.

Thunderstorms are the most common type of severe weather. There may be 40,000 thunderstorms around the world on a single day.

Alexey Stiop/Alamy

Weather Events - Tornadoes

A **tornado** is a severe storm with rotating winds that forms over land. Tornadoes form when small areas of low pressure between the cloud and the ground lift columns of air upward. These areas begin to rotate or spin. Tornadoes can destroy buildings and pick up items as heavy as a truck. The force of the winds can throw the items hundreds of meters away. The paths of some tornados are narrow. Several buildings on a street may be damaged even though others are untouched. Other tornadoes may be 1.6 kilometers (1 mile) wide.

Tornado Safety If you see a tornado or a tornado warning is announced, go to the basement of a building right away. You may also go to an inside room of a building where there are no windows.

Tornadoes, sometimes called twisters, can travel more than 160 km (100 mi). Tornadoes may look like a tall funnel.

Derechos

A *derecho* is a strong windstorm that travels straight over a long distance. A derecho may happen when a series of thunderstorms form over a large area. The winds of a derecho can reach speeds of over 160 km (100 mi) per hour. These storms can be a few miles wide and hundreds of miles long.

Derechoes form most often in the late spring and when the air is warm. The storms may last for hours. They can severely damage trees, homes, and power lines.

Derecho Safety Derechos bring strong, damaging winds. To prepare, people may tape or board up windows. They can make sure nothing outside can blow away or cause further damage. They can also gather flashlights and other emergency supplies in case winds damage power lines.

Word Study

Derecho means "straight" in Spanish. These windstorms form long, straight lines of wind damage.

The straight-line winds of a derecho cause heavy damage.

Weather Events - Hurricanes

A **hurricane** is a large storm with strong winds and heavy rains. A hurricane starts as a storm above warm ocean water. Rising warm air causes the storm to rotate. When winds reach 119 km (74 mi) per hour, the storm is called a hurricane. Some of the strongest hurricanes can reach wind speeds of over 249 km (155 mi) per hour.

Hurricane clouds rotate around a calm area near the center, called an eye. When a hurricane hits land, it can bring intense winds and cause flooding. Hurricanes weaken over land. The warm waters that fed the storm are left behind.

Hurricane Safety Scientists track hurricanes as they form and move. People may be asked to prepare to leave the area to stay safe. Preparing for a hurricane is similar to preparing for a derecho. People prepare for heavy rains and strong winds.

Did You Know?

Hurricanes form over the Atlantic Ocean and the eastern Pacific Ocean. These same types of storms that form over other oceans around the world are called *tropical cyclones.*

Hurricane winds can cause great damage.

This picture shows what a hurricane looks like from space.

Winter Storms

Winter storms occur when a line of warm, moist air is lifted over colder, drier air. Snowstorms, blizzards, and ice storms are all winter storms.

A snowstorm is a winter storm that drops heavy snow on an area. A **blizzard** is a snowstorm with very strong winds. Blowing snow makes it difficult to see. Blizzards bury plants, cars, roads, and buildings under snow. They can cause power outages because of the high winds and heavy snowfall on power lines.

Ice storms cover all surfaces with a slick layer of ice. The ice can be a thin coating or a layer several inches thick. Power lines and tree branches often snap as ice builds up during these storms.

Winter Storm Safety Staying inside and off roads is important during winter storms. Listen to news reports. Roads and airports may close during these storms as snow collects on the ground. After a storm people are advised to remain off roads until they are cleared.

Wally Bauman/Alamy

Earth's History

Fossils

A **fossil** is a trace or the remains of a living thing that died long ago. Fossils can be shells, bones, skin, leaves, or even footprints. They can form in several different ways.

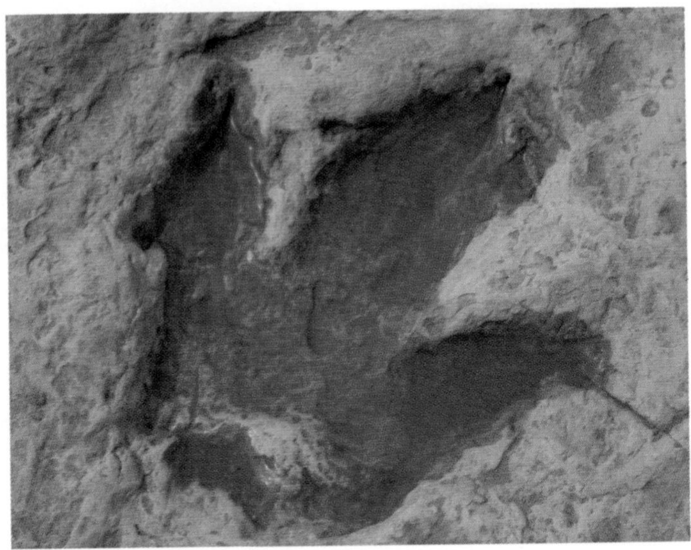

This dinosaur footprint was left in mud. The mud hardened into rock, fossilizing the footprint.

Trace Fossils

Sometimes an animal leaves a footprint on the ground as it walks. The shape of the animal's footprint leaves a mark, or imprint, in mud or clay. The mud hardens and, in time, changes to rock. These fossils are called *trace fossils*. They record a trace of a once living thing.

Preserved Remains

Some fossils are the actual remains of an organism trapped in Earth materials. Amber fossils formed when insects became trapped in tree sap, which hardened over time. Preserved remains also have been found in tar and ice.

This insect's entire body was trapped in tree sap and is now a fossil.

Molds and Casts

A seashell can make an imprint in mud or sand. Over time, the shell is buried in more layers of sand or mud. Slowly, water and minerals seep into the ground. The shell breaks down, but the material around it hardens into rock. The shell-shaped space that is left is called a *mold*. The empty shape can be filled with a new material to show what the original shell looked like. This kind of model is called a **cast**.

The fossil on the right is a mold. The fossil on the left is the cast.

Fossils in Stone

Some fossils, such as bones and teeth, look like the actual parts of animals. When scientists find a bone of a dinosaur, they really have found rock hardened into the shape of a bone. First, sediment buries an animal after it dies. The soft parts of the animal decay. Minerals replace the bones that are left. The rock is the same shape and size as the original animal's body parts.

Fossils can look exactly like the hard parts of the animals they came from.

What Could I Be? Paleontologist

Fascinated by organisms that lived long ago? Paleontologists study fossils of living things of the past. They also work with companies that find fossils while constructing highways and buildings or digging wells. Learn more about the work of paleontologists by turning to the Careers section.

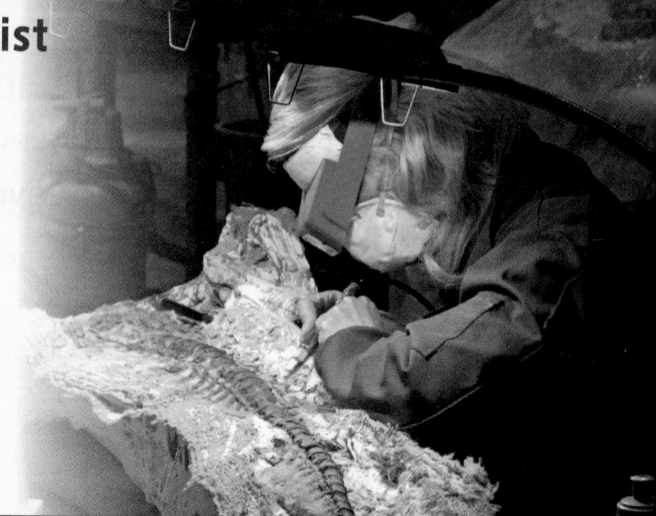

What Fossils Tell Us

Earth is about 4.5 billion years old. People have lived on Earth only a small part of that time. Scientists learn about Earth's past by studying fossils. The *fossil record* is the information about Earth's history found in fossils.

This picture shows fossil trilobites. From their fossils, scientists know that they once lived in oceans all over the world.

We know about organisms that lived long ago because of the fossil record. Trilobites are examples of such animals. No one has ever seen a trilobite, but we know about them because people have studied their fossils.

Changes in Living Things

The fossil record shows that the kinds of things living on Earth have changed over time. Early in Earth's history, many fish swam in the oceans. There were few animals on land. Later, many species of fish died off. **Extinction** is the death of all of one type of living thing. Plants covered the land. Huge fern trees grew. Large insects became common. Changes like these have happened many times.

This fish lived in Earth's seas long ago. It is now extinct.

(t) Siede Preis/Getty Images; (b) National Geographic RF/Getty Images

Evidence of Earth Changes

Looking at fossils also tells us about how Earth's environment has changed over time. Today, Antarctica is a cold place. It is covered in snow and ice. Scientists have found fossils of leaves and wood there. These fossils tell scientists that Antarctica was once a warm, wet area.

Fossil bed at Falls of the Ohio State Park

Today, the states of Indiana and Kentucky have cold winters. Trees lose their leaves as winter approaches. Scientists have found a fossil coral reef in this area. It shows that long ago this area was covered by a shallow, tropical ocean.

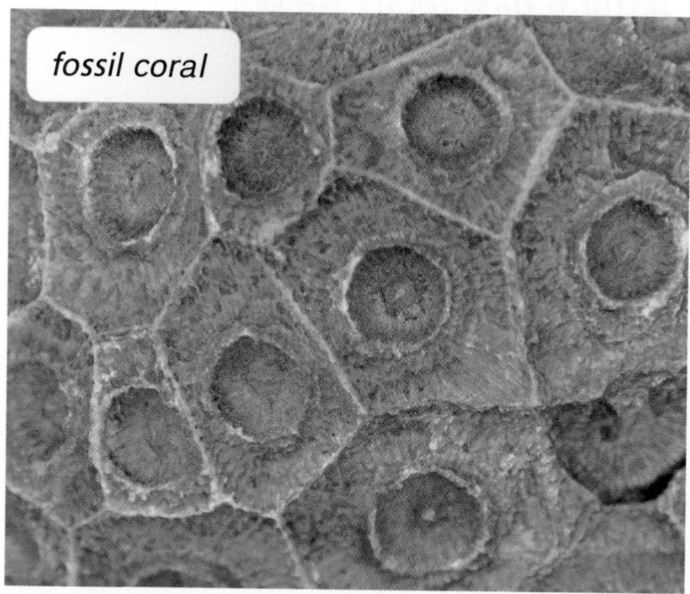

fossil coral

This coral once lived in a warm sea. Over millions of years, the area has changed to dry land in the center of the continent.

Scientists use details they find from fossils to piece together the story of Earth's past. From each new fossil found in Earth's surface, we learn a little more about our planet's history.

Fossil Fuels

People use energy to heat their homes and power their cars. A fuel is a material that is burned for its energy. Coal, oil, and natural gas are called fossil fuels. **Fossil fuels** formed inside Earth from the remains of ancient living things. The remains decayed under pressure for millions of years and remained underground for centuries. Then people started mining coal and drilling for oil and natural gas. Today we burn huge amounts of fossil fuels.

Coal

Coal formed from the remains of ancient plants. The plants lived in swamps. As plants died, they piled up in thick layers. The layers of plants formed *peat*. The peat was buried under sediment that turned into sedimentary rock. Slowly, the peat changed into the sedimentary rock coal.

How Coal Forms

① *Millions of years ago, swamps covered large parts of Earth's land. Over time, the swamp plants died.*

② *Layers of decayed plants formed a fuel called* **peat**. *Then the peat was buried under sediment.*

③ *The sediment turned into sedimentary rock. Slowly the peat changed into the sedimentary rock* coal.

Oil and Natural Gas

Oil and gas did not form from plants. They formed from tiny living things in the ocean. Over millions of years, these tiny organisms died and sank to the ocean floor. The organisms and sediments formed thick layers. Over time, heat and pressure turned this mix into oil. More heat and pressure turn the oil into natural gas.

This rig pumps oil to the surface from the ocean floor.

Thick layers of coal are mined with huge machines.

Did You Know?

There are more than 2.1 million miles of underground gas pipelines across the United States.

Space

When you look out at the sky at night, everything you see is part of space. We use the word *space* to refer to everything we can see when we gaze into the night sky and beyond. There are objects in the night sky that are Earth's neighbors in space.

The Solar System

The Sun, Earth, and Moon are part of a system in space called the solar system. A *solar system* is made up of a star and the objects that move around that star. The star in our solar system is the Sun.

A *planet* is a large body of rock or gas with a nearly round shape. There are eight planets in our solar system. Earth is just one of those planets. The planets are similar in many ways. All of them move around the Sun. All of them *rotate,* or spin, like Earth does.

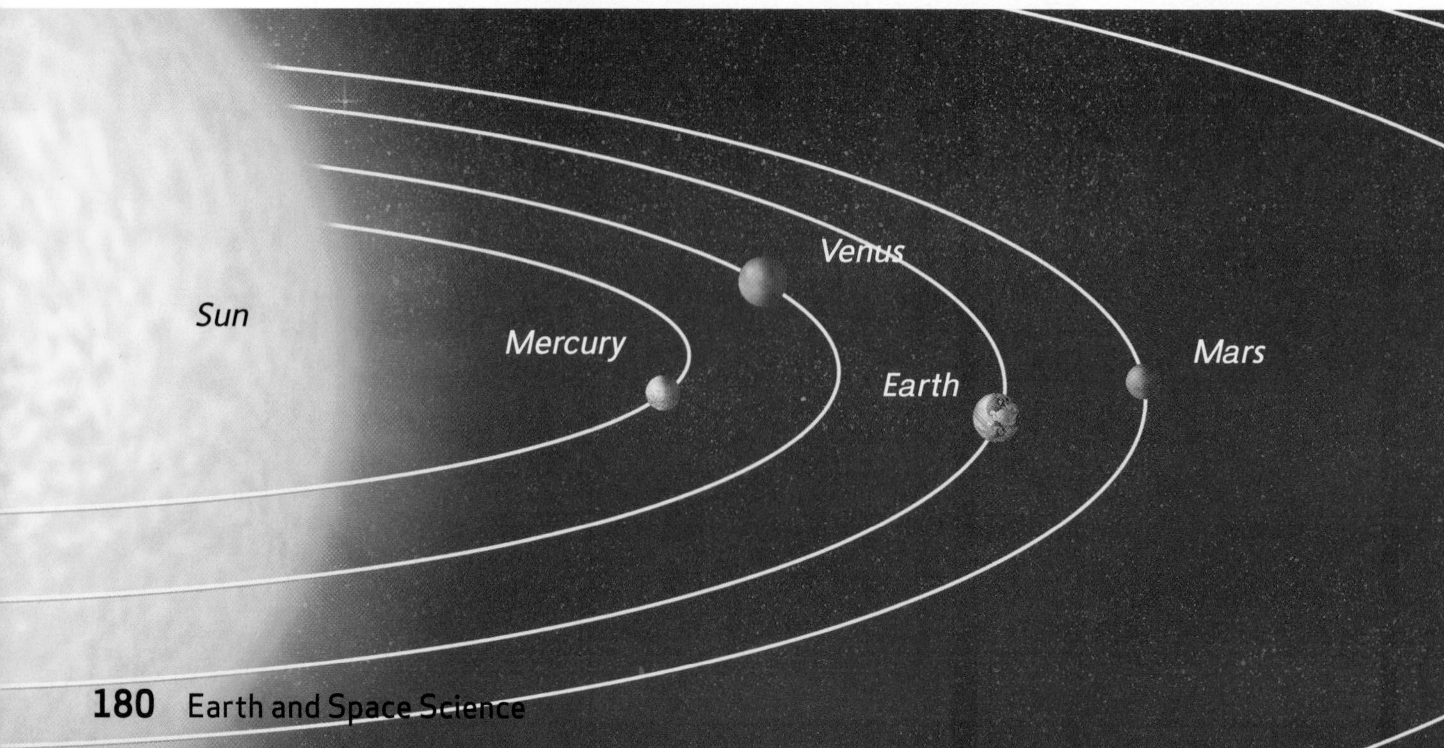

It takes Earth 365 days to go around the Sun one full time. This movement is called an *orbit*. Mercury is the closest planet to the Sun. It orbits the Sun in only 88 Earth days. Neptune is the farthest planet from the Sun. It takes 165 Earth years to make a full orbit.

Seeing Planets at Night

If you look carefully at the night sky, you may see planets. Mercury, Venus, Mars, Jupiter, and Saturn are close enough to see from Earth. The planets look like stars in the sky. Like Earth, however, the planets do not make their own light. We see them because sunlight reflects, or bounces, off them.

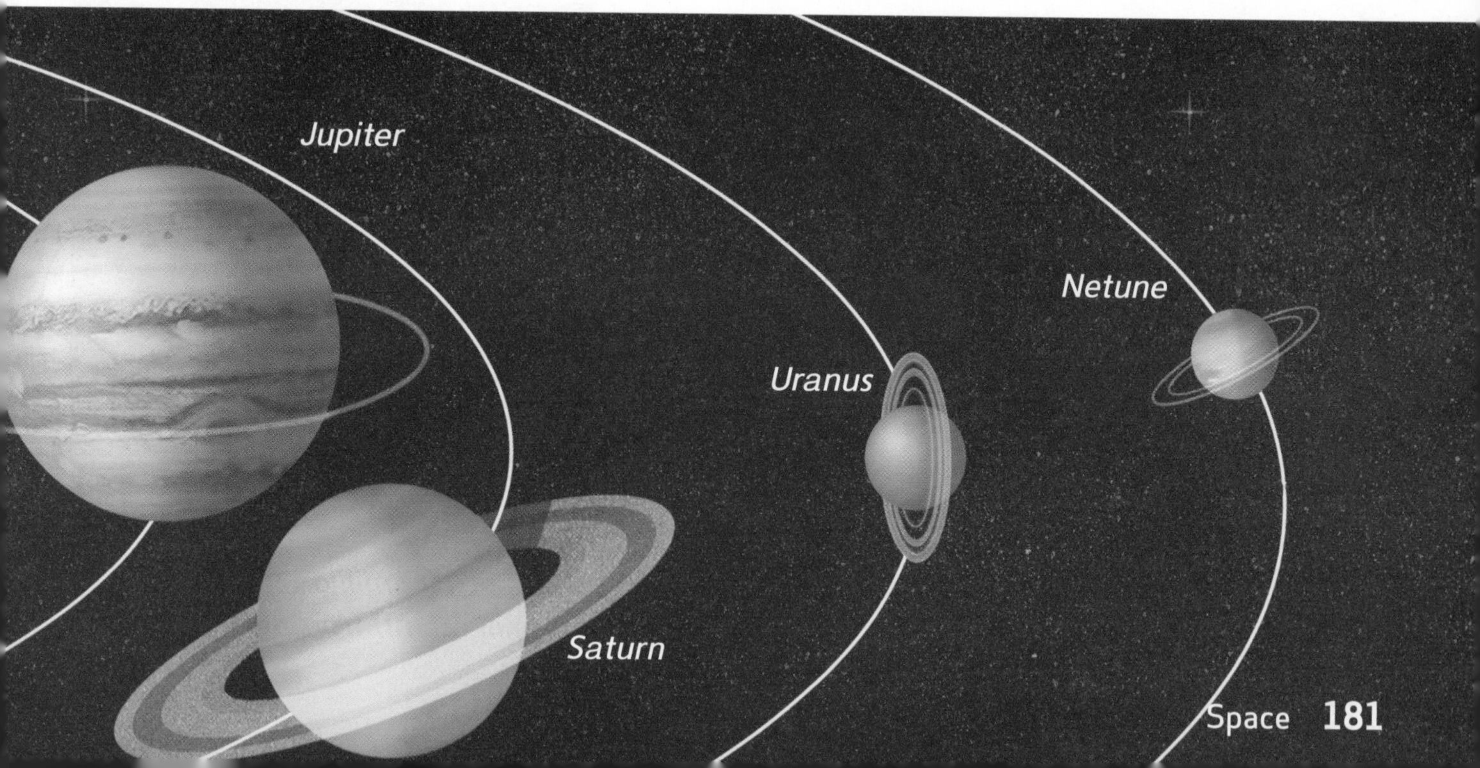

At rare times, several planets can be seen at once in the night sky. This diagram shows how the planets looked in the sky in April of 2002.

The Inner Planets

The solar system's eight planets are divided into two groups. The four planets closest to the Sun are known as the *inner planets.* They are made of solid, rocky materials. The inner planets are the smallest in the solar system. Mercury, Venus, Earth, and Mars are warmer than the other planets because they are closer to the Sun's extreme heat.

Did You Know?

All planets, except for Earth, are named after Greek or Roman gods and goddesses. Mercury is the Roman god of communication and travelers and Venus is the Roman goddess of love and beauty.

Mercury

Mercury is the smallest planet. Even though it is the closest planet to the Sun, Mercury is still 46 million kilometers (29 million miles) from the Sun. The planet is rocky and full of craters. A *crater* is a large hole in the ground.

Mercury has more craters than any other planet in the solar system.

Venus

Venus is the second planet from the Sun. Although it is not as close to the Sun as Mercury, Venus is the hottest planet. That is because it is covered in thick clouds that trap heat.

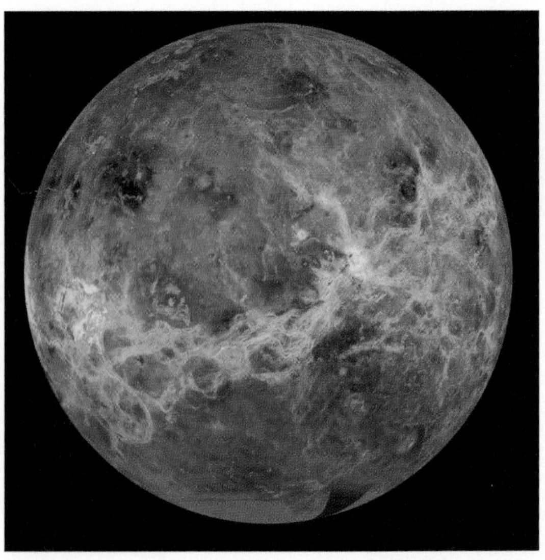

Venus is 107 million km (66 million mi) from the Sun.

Earth

Earth is the third planet from the Sun. It is the only planet known to have oxygen, liquid water, and living things. Earth is 147 million km (91 million mi) from the Sun. Earth is the largest of the inner planets. *Moons* are rocky objects that orbit around a planet. Earth is the only planet with just one moon. Mercury and Venus do not have any moons. The other planets have more than one.

Earth is known as the Blue Planet. Its large amounts of water can be seen from space.

Mars

Mars is known as the Red Planet. It gets this nickname from its reddish-brown soil. When viewed from Earth, Mars has a red glow. Large dust storms are common on Mars. The storms may last for months and may cover the whole planet. A large volcano found on Mars is also the tallest mountain in the solar system.

Mars has polar ice caps that contain frozen water.

The Outer Planets

The four planets farthest from the Sun are called the *outer planets*. Jupiter, Saturn, Uranus, and Neptune are the outer planets. They are called "gas giants" because they are made mainly of gases. They are also larger than the inner planets. Scientists think the outer planets may have solid cores deep below the layers of gases.

Jupiter

Jupiter is the solar system's largest planet. It has two large red spots on its surface. One of these is called the Great Red Spot. It is a huge storm system. Scientists think the storm may have been blowing for over 350 years. Compared with Earth, days on Jupiter are short. Earth spins once every 24 hours. Jupiter spins once every 9 hours and 55 minutes.

Jupiter is the planet with the most mass. It is more than twice the mass of all of the other planets combined.

Saturn

Saturn is known for its large rings. The rings are made of bits of ice and rock. The rings revolve, or orbit, around the planet. Saturn has about 150 frozen moons.

Saturn's rings were first discovered by the scientist Galileo in 1610.

Uranus

Uranus is the seventh planet from the Sun. It takes 84 Earth years for Uranus to make just one orbit around the Sun. Uranus moves a bit differently than other planets. It is called the sideways planet because it rotates on its side. It is the coldest planet in the solar system. Like Saturn, Uranus has rings around it. Uranus was the first planet discovered using a telescope.

Uranus was named after the Greek god of the sky. It is the only planet named after a Greek god.

Neptune

At a distance of about 4.5 billion km (2.8 billion mi), Neptune is the farthest planet from the Sun. The planet has a very thick atmosphere of gases. These gases give the planet its blue appearance. They also cause large storms. One storm, called the Great Dark Spot, lasted about five years. Neptune has not been investigated as much as the other planets. It is far away and difficult to observe.

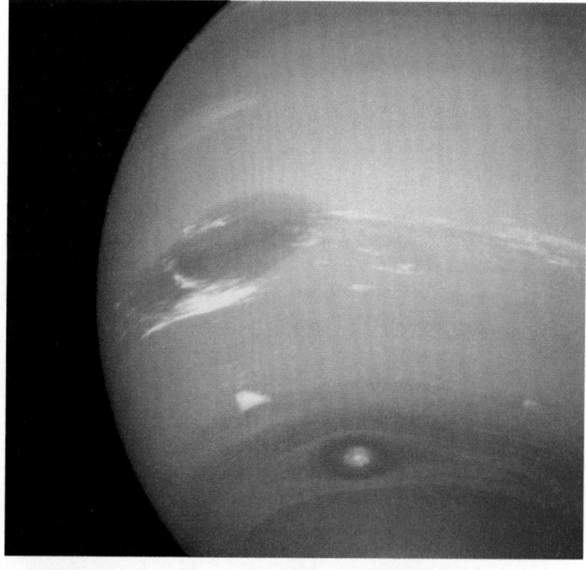

Earth gets its blue color from surface water, but Neptune gets its blue color from gases.

Earth in Space

Earth rotates, or spins, like a giant top in space. As Earth rotates, different parts of the planet have day and night.

Day and Night

At any one time, the part of Earth that faces the Sun has daytime. At the same time, the other side of Earth has nighttime. That side of Earth is facing away from the Sun.

Earth keeps spinning. Day becomes night for one side of Earth. Night becomes day for the other side of Earth. This pattern happens again and again. It takes about 24 hours for Earth to make one complete turn on its axis. An **axis** is a line through the center of a spinning object. One complete turn is one day.

Fact Checker

A day is not just daytime hours. A day is made up of daytime and nighttime hours.

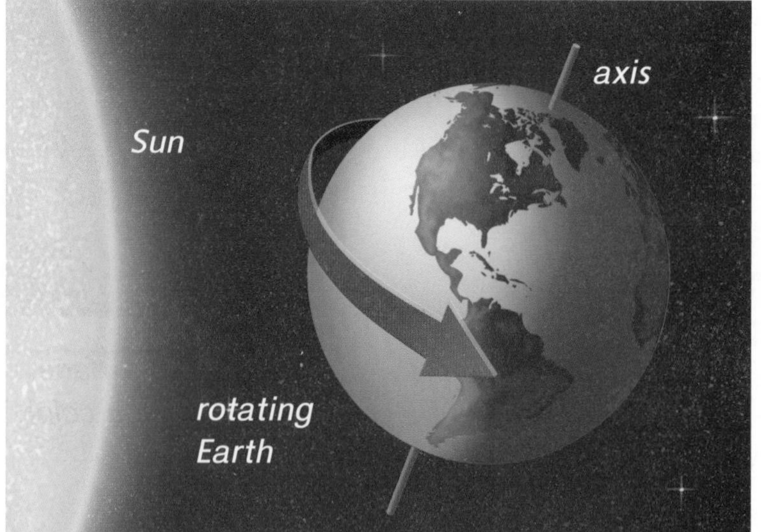

As Earth rotates on an imaginary line called an axis, we have day and night.

The Sun in the Sky

As Earth rotates, the Sun appears to move across the daytime sky. Early in the morning, the Sun looks like it is rising. It looks low in the eastern sky. The Sun's light casts long shadows when it is low in the sky. At midday, the Sun is high in the sky. We see short shadows on the ground at midday. In the evening, the Sun looks like it is setting in the sky. It looks low in the western sky. The Sun's light casts long shadows again. The shadows point in a direction opposite the morning shadows.

This changing position of the Sun in the sky makes it look as if the Sun is moving. It is not moving. Earth is moving. Earth is spinning on its axis.

The Sun seems to rise in the east and set in the west. These pictures are drawn as if you are facing south.

Earth in Space - Seasons

At the same time Earth spins, it also revolves around the Sun. To *revolve* means to move around another object. One full movement of an object in a regular path is called an *orbit.* It takes Earth about 365 days to orbit the Sun. This length of time is called one year.

Earth's axis is not straight up and down. It is on a tilt and always points in the same direction. As Earth orbits the Sun on its tilted axis, we experience different seasons. For part of Earth's orbit, the upper half of Earth, called the *Northern Hemisphere*, is tilted toward the Sun. It gets more direct rays of energy from the Sun. The Northern Hemisphere has summer. At the same time, the *Southern Hemisphere,* or lower half of Earth, is tilted away from the sun. The Southern Hemisphere has winter.

Skill Builder

Read a Diagram

Look carefully at Earth's axis. It stays in the same position as Earth moves around the Sun. The circular arrows around the axis show that Earth is rotating. The large yellow arrows show that Earth is revolving around the Sun.

The labels name the seasons in the Northern Hemisphere. Also think about what season the Southern Hemisphere is having in each picture.

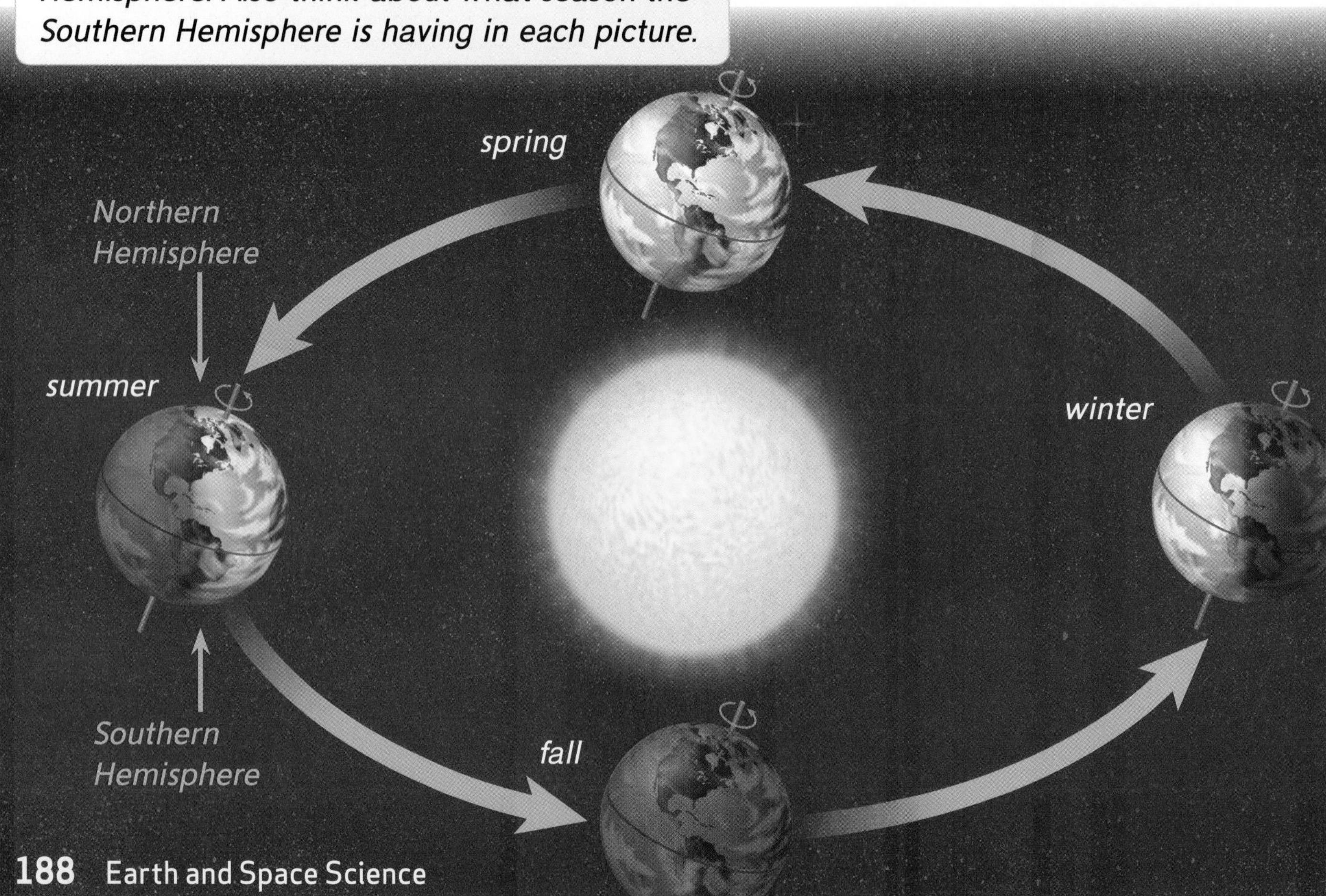

At other times, the Northern Hemisphere is tilted away from the Sun and has winter. At those times, the Southern Hemisphere is tilted toward the Sun and has summer.

The Northern Hemisphere gets the most light and heat from the Sun in summer. There are many hours of daylight each day. Temperatures are warmer than in any other season. As Earth continues in its orbit, temperatures in the Northern Hemisphere cool. There are fewer daylight hours. Fall begins.

The Northern Hemisphere gets the least light and heat from the Sun in winter. There are fewer hours of daylight. Temperatures are cooler than in other seasons. As Earth continues in its orbit, the number of daylight hours increases. Temperatures get warmer. Spring begins.

The seasons follow a pattern. We can predict when each season will begin and end. We can predict the amount of sunlight and the temperatures of each season.

Each season brings changes to living things. Many trees change based on the temperature and the amount of sunlight they receive.

The Moon

The Moon is Earth's closest neighbor in space. It is a sphere, or ball, of rock. The Moon is a satellite of Earth. A *satellite* is an object that revolves around another larger object in space. The Moon revolves around Earth, much like Earth revolves around the Sun.

When the Moon looks like a big circle, it is in full moon phase.

Like the Sun, the Moon appears to rise, move across the sky, and set. This movements seems to happen because of Earth's rotation. As Earth rotates on its axis, the Moon's position in the sky moves from east to west.

The Moon seems to glow in the night sky. However, the Moon does not give off its own light. The light we see comes from the Sun. Sunlight reflects, or bounces, off the Moon's surface.

Moon Phases

Phases of the Moon

Each day the moon seems to change shape. Sometimes you can see only a small part of the Moon. At other times, it is a big, bright circle. At still other times, you cannot see the Moon at all. Each shape of the Moon we see from Earth is called a *phase.* The diagram shows the eight main phases of the Moon. This pattern repeats every 29½ days.

The Moon does not reflect sunlight only at night. Sometimes we can see the Moon during the day.

Skill Builder **Read a Diagram**

Notice how the part of the Moon we can see from Earth changes. The lit area visible to us grows larger, and then smaller.

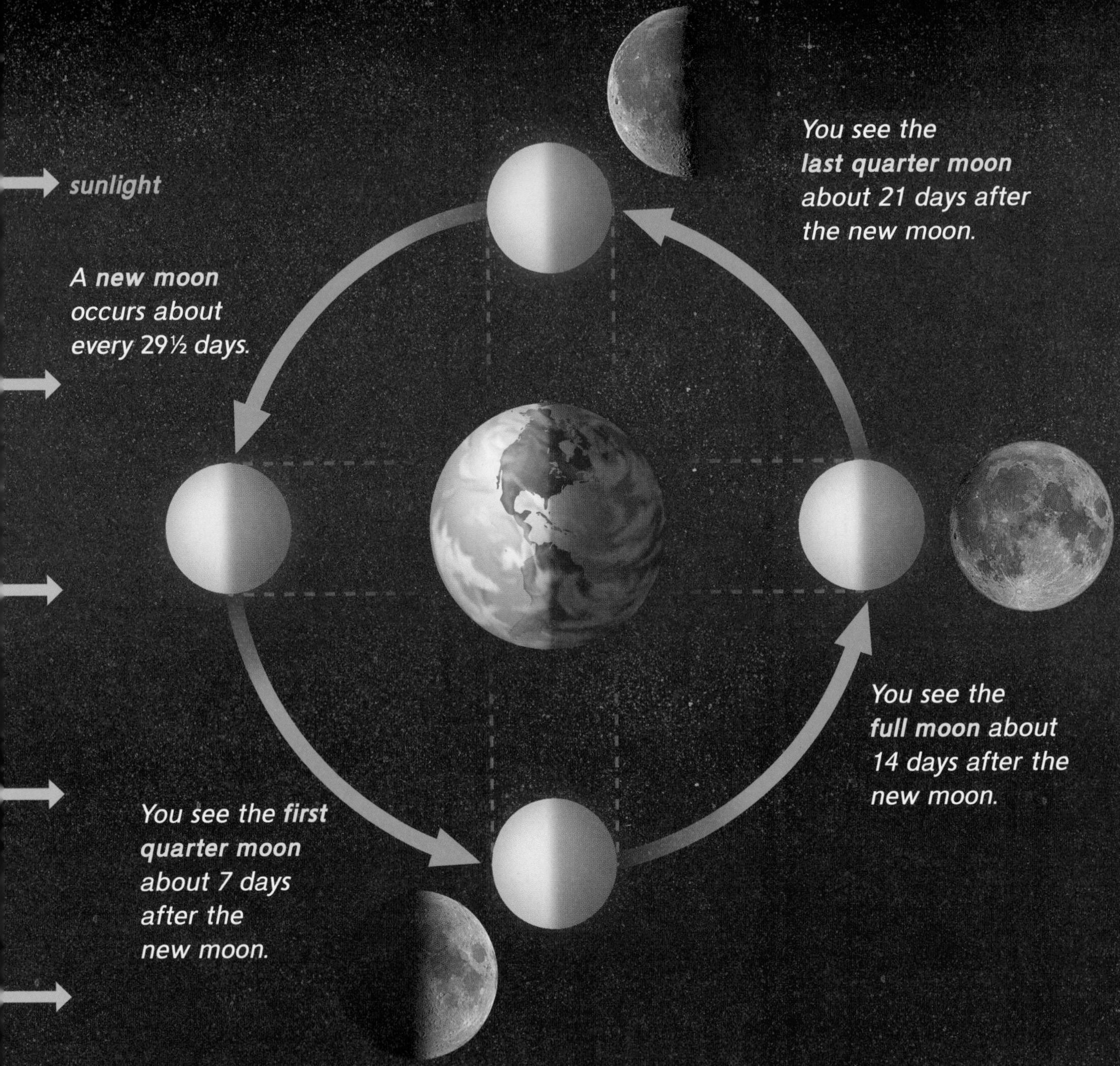

sunlight

A new moon occurs about every 29½ days.

You see the last quarter moon about 21 days after the new moon.

You see the full moon about 14 days after the new moon.

You see the first quarter moon about 7 days after the new moon.

The photos show what the first quarter moon, full moon, and last quarter moon look like from Earth.

Skill Builder **Read a Diagram**

Notice the difference in the arrows in this diagram. The arrows on the left indicate light from the Sun. The arrows around Earth indicate the Moon's path.

The Moon - Causes of Moon Phases

The Moon is a sphere. It does not change shape. It appears to change shape because the amount of its sunlit side that we can see changes. Just like Earth, half of the Moon is in light while the other half is in darkness. This fact is true all the time. As the Moon revolves around Earth, a different amount of the Moon's sunlit half faces Earth. The lit parts are the phases you see.

When the Moon is between Earth and the Sun, there is a new moon. You cannot see the Moon at all. The Moon's lit side faces away from Earth. As the Moon continues to revolve, you see more and more of the Moon's lit side. About two weeks after the new moon, you can see the entire lit half of the Moon. This phase is a full moon. The Moon continues its journey. Less and less of the Moon's lit side is visible. About every 29½ days, the Moon returns to the same position. The cycle repeats again and again.

The surface of the Moon has many craters.

Comparing Earth and the Moon

In some ways, Earth and the Moon are the same. They both are rocky spheres. They both rotate and revolve. They both get light from the Sun.

In other ways, Earth and the Moon are different. Earth is about four times larger than the Moon. The surface of the Moon is covered with craters, or pits. The craters formed when chunks of rocks from space crashed into the Moon. There is no atmosphere on the Moon. There is no liquid water. Nothing lives on the Moon.

Stars

A *star* is a ball of hot, glowing gases. On a dark, clear night, millions of stars are visible.

The Sun

The Sun is the star in the center of our solar system. In fact, it is the only star in our solar system. It is a medium-sized star. The Sun looks big and bright because it is the star closest to Earth. Because it is so bright, we cannot see any other stars during daylight.

The Sun is so hot that it could melt Earth. Inside the Sun, temperatures reach about 15 million degrees Celsius (59 million degrees Fahrenheit). However, the Sun is 150 million km (93 million mi) away from Earth. It would take a person 160 years to travel that distance by car. The amount of heat and light we get from the Sun is enough to sustain life on Earth. Without the Sun's heat and light, plants could not grow. Earth would freeze.

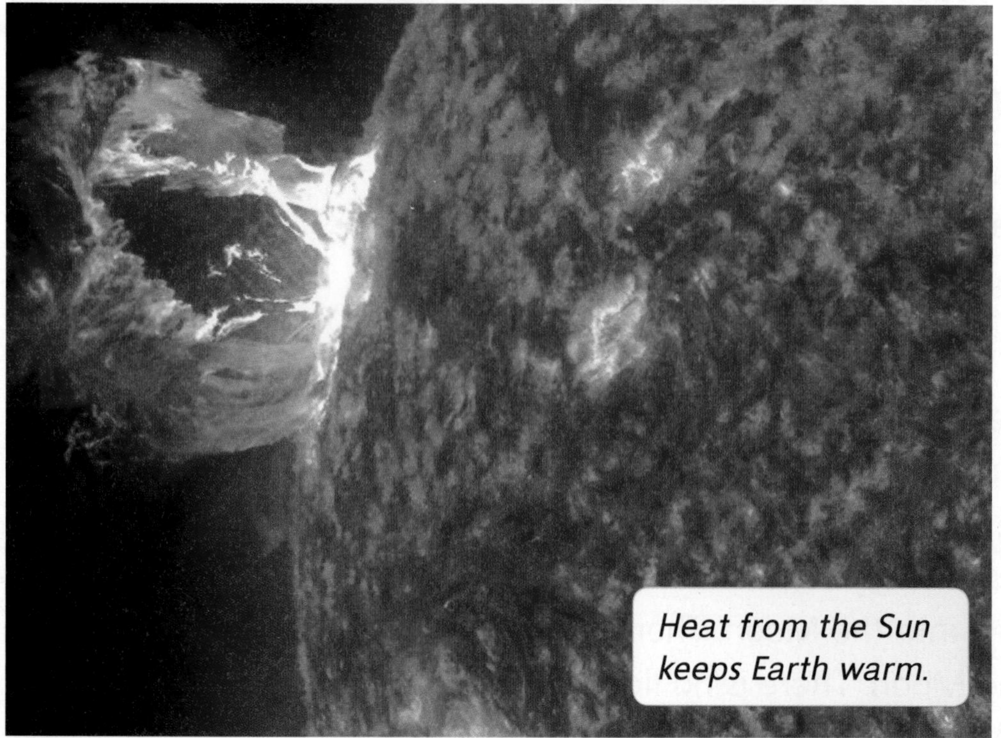

Heat from the Sun keeps Earth warm.

NASA/SDO

Other Stars

The sky is filled with more stars than can be counted easily. Stars are always in the sky, but the Sun's light prevents us from seeing them during the day. All the stars we see in the night sky are much farther away from Earth than the Sun is. Some stars seem bigger and brighter than others. That is because they are either closer to Earth or they give off much more light energy than other stars. Some stars seem tiny and faint because they are so far away or they do not give off as much light as others.

Stars do come in very different sizes, however. Large stars may be several times bigger than the Sun. Small stars may seem like a small dot compared to the Sun. Some giant stars are thousands of times larger than the Sun.

Color and brightness are other ways that we can describe stars. Blue stars are the hottest. Red stars are the coolest.

Stars differ in size, color, and brightness.

Stars - Constellations

People have gazed at the night sky for thousands of years. Long ago, people thought that groups of stars formed pictures in the night sky. The pictures they described can still be seen today. A group of stars that seem to form a picture is called a *constellation*.

A noticeable pattern in the night sky of the Northern Hemisphere is the Big Dipper. It looks like a drinking cup with a long handle, but is not a constellation by itself. The Big Dipper is made of the seven brightest stars in another constellation. This larger constellation is called the Great Bear, or Ursa Major. The stars form the head, body, legs, and tail of a bear. The handle of the Big Dipper is also the tail of the Great Bear.

The constellations were named long ago. Today people might see different pictures if they were to connect stars in the sky.

The Big Dipper is part of the constellation Ursa Major. The yellow lines show the stars that make up the Big Dipper.

Space Exploration

People have always wondered about objects in the night sky. There is still a lot to learn about space. Some tools help us see space from Earth. Other tools can be sent into space to collect information and send data back to Earth.

Telescopes

A **telescope** is a tool that makes faraway objects appear closer and larger. Mirrors and lenses in the telescope gather and focus the light. A *refracting telescope* uses a curved piece of glass, called a lens, to bend light. A lens makes distant objects seem larger. The first telescopes used lenses. A *reflecting telescope* uses mirrors to reflect light and make an object seem larger. The Hubble Space Telescope that orbits Earth is a reflecting telescope.

Galileo was the first scientist to use a telescope to view the night sky. Today telescopes are much stronger. They can see farther into space with more detail.

What Could I Be?

Astronomer

Interested in the planets and space? Think about a career as an astronomer! These scientists use telescopes and other tools to study objects in our solar system and beyond. They use computers to analyze the data they collect. Learn more about what astronomers do by turning to the Careers section.

Space Exploration - Space Probes

A *space probe* is a machine that is sent from Earth into space to collect information. A probe may take pictures or gather data. Probes send data back to Earth for scientists to study.

Voyager 1 is a probe that was launched in 1977. The probe was made to gather data on Saturn and Jupiter. It discovered active volcanoes on one of Jupiter's moons. It gathered data on Saturn's rings. It then headed for the edge of our solar system. The probe is now over 19 billion km (11 billion mi) from Earth. It has traveled farther in space than any human-made object!

Scientists have sent two rovers to Mars to find out about water on the planet. They also search for and analyze rocks and soils. The rovers carry cameras, magnets, temperature sensors, and other tools.

Voayger 1 took this picture of Saturn's rings in 1980.

One of the rovers has traveled over 40 km (25 mi) across the surface of Mars.

The International Space Station

Humans are learning more about space every day. The *International Space Station (ISS)* is a large science laboratory in space. It is a place where astronauts from around the world live and work together. The space station orbits about 354 km (220 mi) above the ground.

The space station helps scientists learn about what happens to people when they live in space. They have learned how to fix a machine in space. The scientists on the ISS do experiments. They take photos of Earth's surface. They see how space affects chemical processes.

Looking Ahead

The International Space Station provides a way to prepare astronauts to explore farther out in space in the future. There is still a lot to learn about our own solar system. However, scientists also hope to some day learn more about what is beyond our solar system.

Six people can live and work on the ISS at one time.

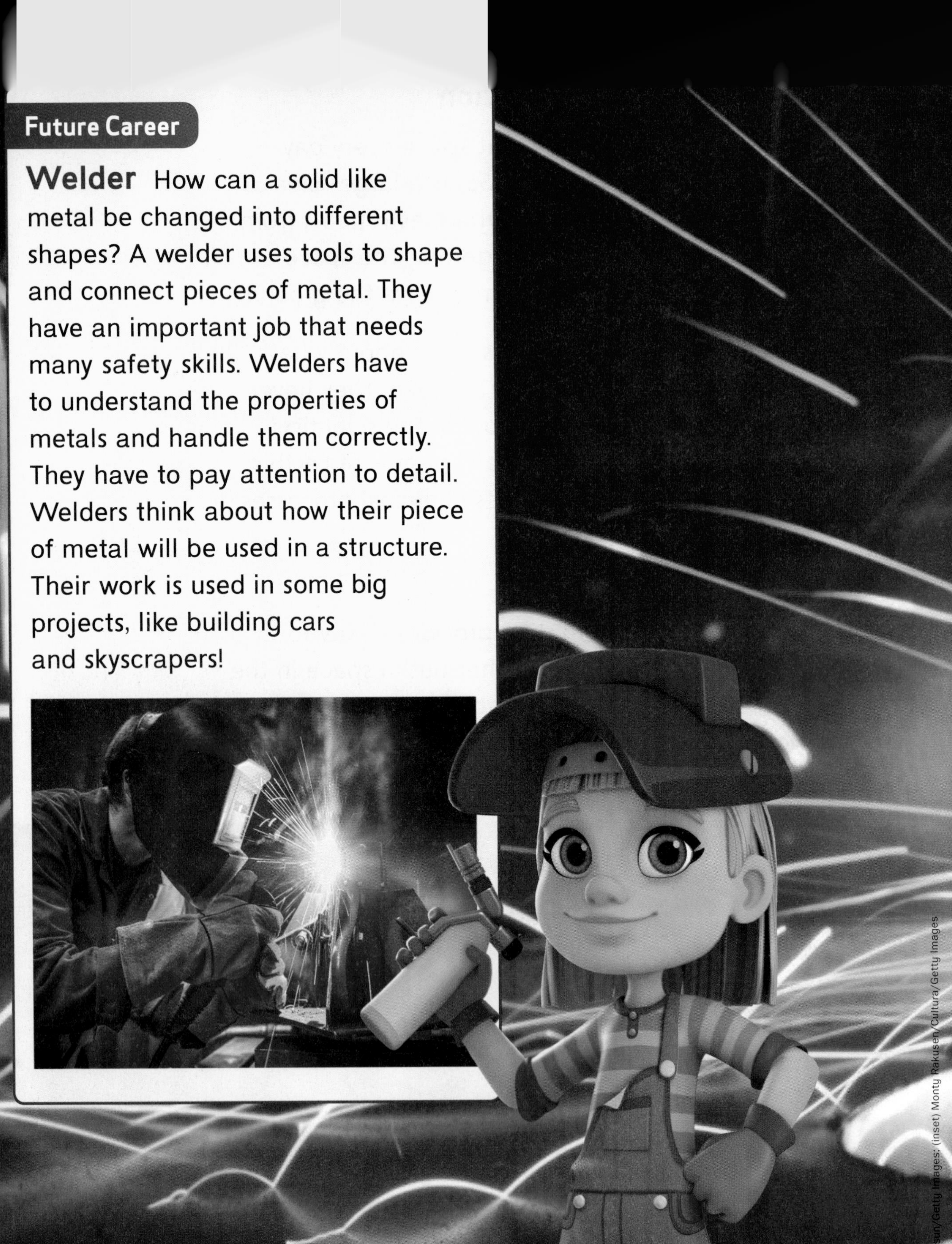

Future Career

Welder How can a solid like metal be changed into different shapes? A welder uses tools to shape and connect pieces of metal. They have an important job that needs many safety skills. Welders have to understand the properties of metals and handle them correctly. They have to pay attention to detail. Welders think about how their piece of metal will be used in a structure. Their work is used in some big projects, like building cars and skyscrapers!

Physical Science

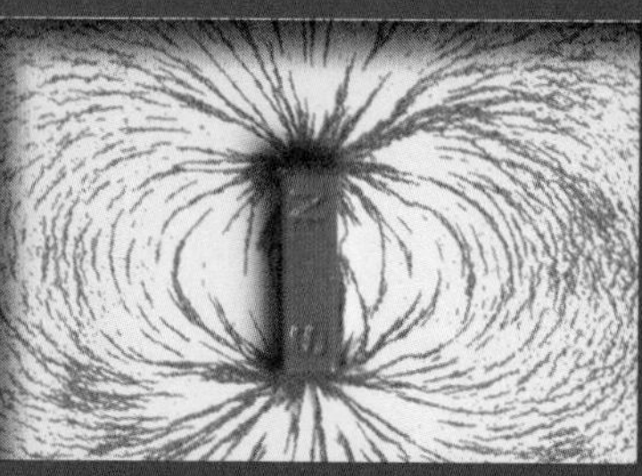

Matter

Defining Matter

During a hurricane, rain pours from the sky. Waves move sand on the shore. Palm trees are bent from the wind. The rain, the sand, the palm trees, and the wind are all matter. *Matter* is anything that takes up space. Since matter takes up space, two objects cannot take up the same space at the same time.

You do not have to be able to see something for it to be made of matter. The air you breathe takes up space, so it is matter.

Properties of Matter

Matter is described by its properties. A *property* is a characteristic of something. The way an object looks, tastes, smells, sounds, and feels are properties you can observe. Some properties can also be measured.

Everything in this picture is made of matter.

FEMA/Tim Burkitt

Physical Properties of Matter

Anything you can observe about matter is a *physical property*.

Volume Because matter takes up space, it has volume. *Volume* is the amount of space something takes up. A beach ball takes up more space than a bowling ball. Because a beach ball is bigger than a bowling ball, a beach ball has greater volume than a bowling ball has.

Mass All objects have mass. *Mass* is a measure of the amount of matter in an object. An object with a large mass, such as a bowling ball, feels heavy. An object with a small mass, such as a beach ball, feels light. The bowling ball has greater mass than the beach ball because the bowling ball contains more matter.

Fact Checker

Bigger objects do not always contain more matter. Mass—not volume—shows how much matter is in an object.

Properties of a Beach Ball and a Bowling Ball

Property	Ball	Description
Color(s)	beach ball	red, green, white, orange, blue, yellow
	bowling ball	red
Shape	beach ball	round
	bowling ball	round
Feel	beach ball	flexible
	bowling ball	hard

Skill Builder

Read a Table

Tables are made of columns and rows. The headings in the first row help you understand the information in the columns below.

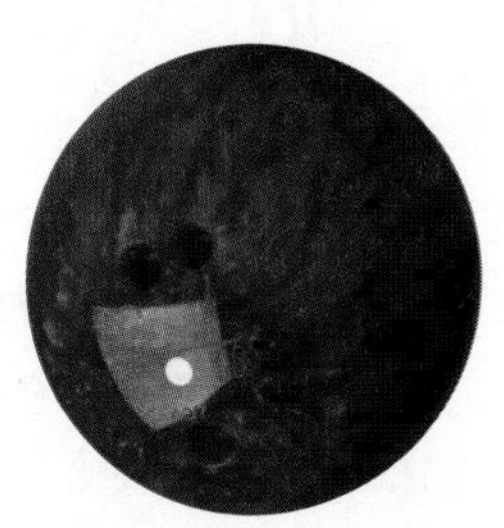

This beach ball has greater volume and less mass than the bowling ball.

Physical Properties of Matter - Strength

A material that can support a heavy load without breaking or tearing is strong. *Strength* is a physical property of matter. Bridges are built to hold up heavy loads. The materials used to build a bridge are chosen for their strength.

The iron and steel used to build a bridge are strong. They will not break under heavy loads.

Flexibility

Some materials can bend and stretch and then return to their original shape without breaking. These materials are said to be flexible. *Flexibility* is a physical property of matter. Rubber bands are flexible. Metals, such as copper, are flexible when they are shaped into a thin sheet or wire.

Magnetism

Magnets pull on, or attract, certain metals such as iron. They do not attract wood, plastic, or water. **Magnetism** is a measure of how much an object is attracted to a magnet. Magnetism is a physical property of iron. It is not a physical property of most other kinds of matter.

These materials are flexible. They bend or stretch without breaking.

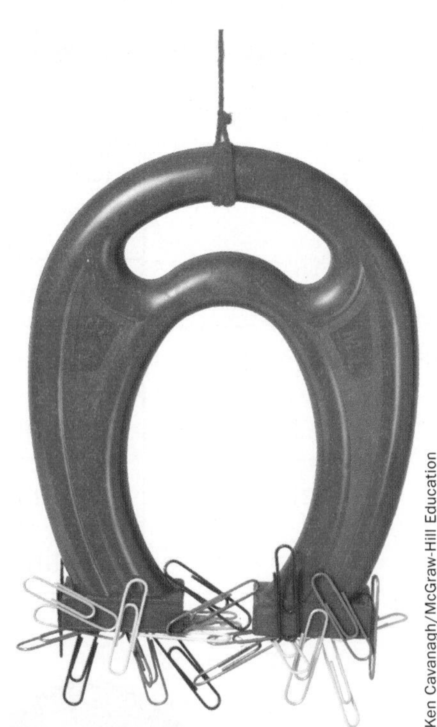

The paperclips are made of steel, which contains iron. They are attracted to the magnet.

Hardness

The *hardness* of a material refers to how easily it can be scratched. Hard materials can scratch softer materials. The chalk used to write on a chalkboard is soft. It is easily scratched off by the harder board, leaving a mark.

Chalk is soft enough to rub off on the harder board.

Texture

Texture is how something feels. Words that describe texture include slimy, rough, fuzzy, oily, and smooth.

The pineapple has a rough texture. The seashell has a smooth texture.

Sinking and Floating

Some matter, such as a rock, sinks in water. Other matter, such as an apple, floats. Metal objects usually sink, and wooden objects often float. Objects sink or float because of their mass and volume. Objects with a lot of mass and a little volume tend to sink. Objects with a little mass and a lot of volume tend to float.

The bobber floats. The hook sinks.

States of Matter

All matter is made of particles. The way those particles are arranged determines the matter's form.

The water is a liquid. The canoe is a solid. The air is a gas.

Solids, liquids, and gases are three forms of matter. Scientists call these forms *states of matter.* Each state of matter has certain properties.

Solids

Most of the things you notice around you are solids. A *solid* is matter that takes up a definite, or certain, amount of space. A solid has its own shape. A book is a solid. Pencils, desks, and pillows are solids, too. Solids can be hard or soft. Solids have a definite shape and volume.

The particles that make up solids are packed closely together. They do not have a lot of room to move around. This movement helps the solid keep its shape.

The particles in this solid kayak can not move much.

(b) Comstock Images/Alamy; (t) ©Brand X Pictures/Superstock

Liquids

A *liquid* is matter that has a definite volume, but not a definite shape. A liquid takes the shape of its container. Water, shampoo, and milk are liquids. Particles in a liquid are more free to move past one another. When you pour a liquid from one container to another, its shape changes as the particles slide around.

The particles in a liquid are able to slide past one another. That is why liquids take the shape of their containers.

Gases

You cannot always see gases, but they are all around you. A *gas* is matter that has no definite shape or volume. The particles in a gas can move about freely. If a gas is moved from one closed container to another, the gas particles will completely fill the new container.

The particles in the helium gas can move about freely. The gas changes shape and volume as it leaves the tank and fills the balloons.

Chemical Properties

A *chemical property* is a characteristic of a substance that describes its ability to change into different substances. You cannot identify chemical properties by looking at or touching an object. The matter must be acted on in some way in order for chemical properties to be identified.

Burning

The ability to burn is a chemical property. Suppose you want to observe or measure properties of a piece of paper. You can observe the color and odor of the paper without changing the kind of matter in the paper. The paper was still paper after observing these properties. But if you want to know whether the paper will burn or not, you have to burn the paper. Burning changes the paper from one substance into new substances—ash and gases.

Paper and wood are materials that burn easily. Other materials, such as gold, do not burn at all.

Oleksiy Maksymenko Photography/Alamy

Oxidizing

Different metals can combine with oxygen and water in the air. This process can turn some strong metals into a soft crumbling powder. On the right, you can see some examples of the ways different materials are affected by the air.

Objects that contain iron often rust if they are left untreated. Painting protects metal from rusting.

When copper oxidizes, it reacts with oxygen and water in the air and turns green. You can see this chemical property on copper used in pennies, buildings, or statues.

Heating

Some kinds of matter change when they are heated. You can observe chemical properties by cooking food. Many properties of an egg change as it is cooked.

What Could I Be? Materials Engineer

Looking for ways to test the physical and chemical properties of different materials? Materials engineers develop and test materials to make all kinds of products. Everything from computer chips to airplane wings depends on the work of these engineers. Learn more about becoming a materials engineer in the Careers section.

Observing and Measuring Matter

Senses can be used to observe physical properties that involve size or amount. You use your ears to hear the sound a guitar string makes when you pluck it. Your senses of taste, sight, and smell can be used to observe the properties of a fresh apple. When you rub the bark of a tree, you observe that it feels rough.

Many properties of matter can also be observed or measured with tools. You can look closely at an object with a hand lens. You can measure its length and width with a ruler. You can use a thermometer to measure its temperature.

Measuring is a way to compare sizes or amounts. People use tools marked with standard units to measure matter. A *standard unit* is a unit of measurement that people agree to use, such as feet or miles. A common system of standard units is the **metric system**. Scientists use the metric system.

Make Connections

Jump to the Science Guide section to learn more about units in the metric system.

Senses can be used to observe the taste, smell, feel, shape, and color of these fruits and vegetables.

Measuring Length and Width

An object's *length* is the number of units that fit from one end to the other. *Width* is the number of units that fit across. A family member may have used a tape measure or a ruler to measure your height. Your height is the length from the floor to the top of your head. In the metric system, length is measured in units called meters. Long distances are measured in kilometers. A kilometer is the same as 1,000 meters.

Measuring Liquid Volume

Recall that volume describes how much space an object takes up. You have probably used measuring cups to measure the volume of liquids. In the metric system, liquid volume is measured in units called liters (L) and milliliters (mL). Common tools for measuring liquid volume include beakers, graduated cylinders, and measuring cups.

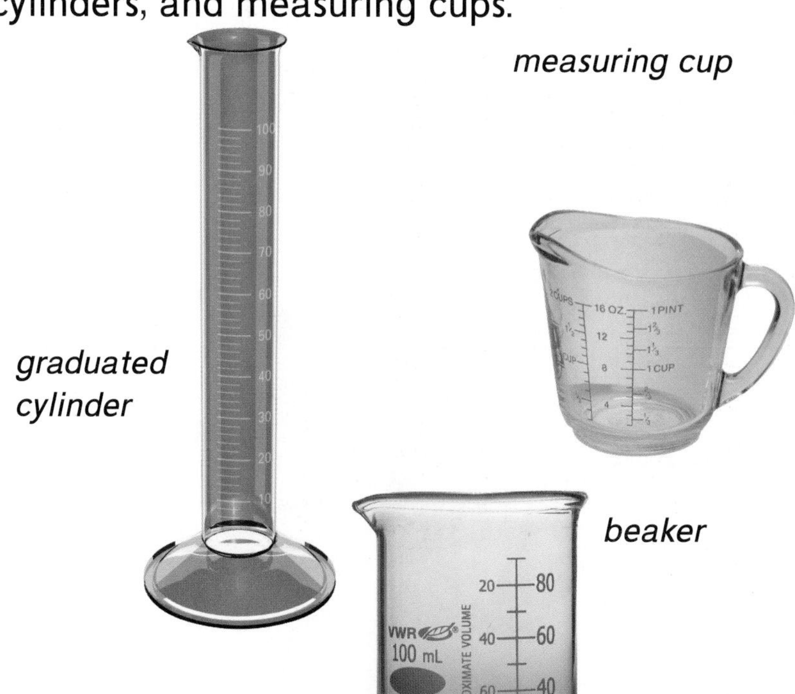

measuring cup

graduated cylinder

beaker

What Could I Be?

Construction Manager

Construction managers plan and direct construction projects. They measure to determine how much material is needed for each job and how much space the materials will fill. Turn to the Careers section to find out more about being a construction manager.

Observing and Measuring Matter - Volume of Solids

In a solid, volume describes the number of cubes that fit inside an object. To find the volume of a rectangular object, multiply its length by its width and height.

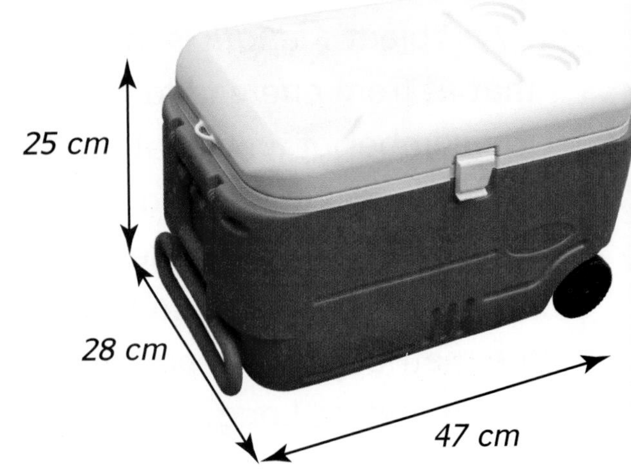

If a solid is not rectangular, you can use water to determine volume. Follow the steps listed in the diagram.

Measuring the Volume of a Solid

Follow these steps:

1. *Measure some water.*
2. *Place a solid object completely under the water.*
3. *Subtract the original water level from the new water level. The difference is the solid's volume.*

Skill Builder **Read a Photo**

Look carefully at the scale on the side of the beaker to understand how to measure the volume of a rock.

(t) Beth Van trees/Hemera/Getty Images; (b) Ken Karp/McGraw-Hill Education

Measuring Mass

Mass is the measure of the amount of matter in an object. Mass can be measured using a pan balance. To find an object's mass, you balance it with objects that have masses you know. First place the object on one pan of the balance. Then add the known masses to the other pan until both pans are level.

Gram masses come in standard sizes.

In the metric system, mass is measured in grams. A gram is close to the amount of mass in two small paper clips. A kilogram is the same as 1,000 grams.

Objects with the same volume do not always have the same mass. Matter is made up of tiny particles. The particles in a marble are packed together more tightly than the particles in a piece of popcorn. A marble has more particles that are closer together than a piece of popcorn. It has more mass.

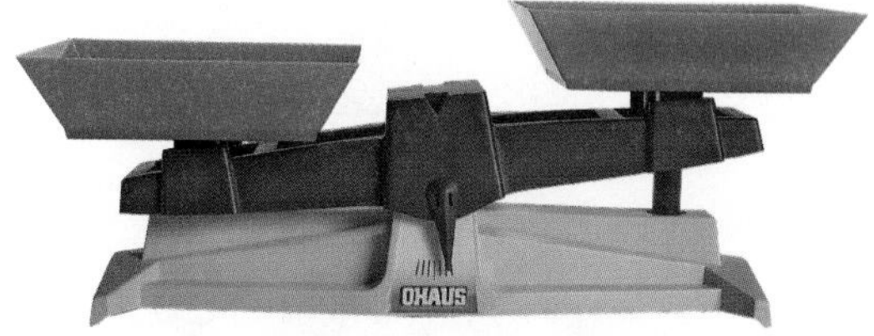

pan balance

The bag of marbles has more matter than the same-sized bag of popcorn. That is why it has more mass.

Changes in Matter

Matter can change. During a *physical change,* the appearance of an object might change, but the kind of matter in the object stays the same. A *chemical change* is a change that causes different kinds of matter to form.

Physical Changes

There are many types of physical changes. Not all types of matter change in the same way. If you pull on a rubber band, it stretches. When you let go, it returns to its original shape. If you pull on a metal spoon, nothing happens. If you pull on a piece of thread, it might break.

Common Physical Changes

If you cut or tear a piece of paper, it looks different, but it is still paper.

If you roll some foil into a ball, it changes shape, but it is still foil.

Even when bent, these wires are still made of plastic-covered copper wire.

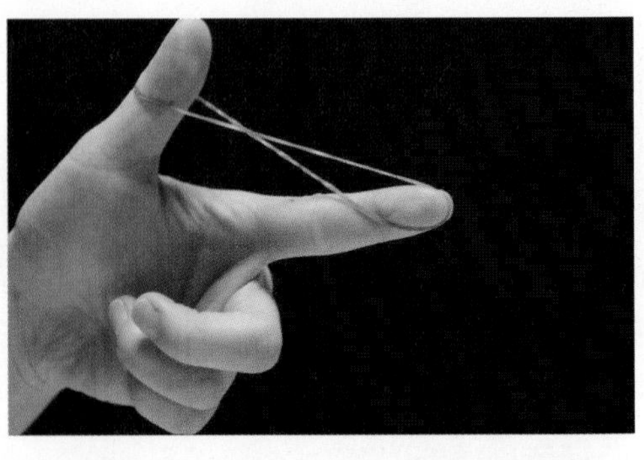

A rubber band changes length and size when it is stretched, but it is still made from rubber.

Changes of State

The *state* of matter means whether an object is a solid, a liquid, or a gas. Changes of state are physical changes. Changes of state are caused by energy changes.

Adding Energy When a solid is heated, it gains energy. Its particles begin to move away from one another. If enough heat is added to most solids, they will melt. To *melt* is to change from a solid to a liquid. Adding heat to a liquid adds energy to the liquid. The particles move faster and farther apart. Eventually, the liquid will boil. When a liquid *boils,* it changes from a liquid to a gas.

Losing Energy When matter is cooled, it loses energy. The particles move more slowly and closer together. If you cool a gas to the right temperature, it will condense. To *condense* is to change from a gas to a liquid. If you keep cooling the liquid material, eventually it will freeze. To *freeze* is to change from a liquid to a solid.

Water Changing State

Boiling water creates bubbles. These bubbles are water in the gas state, or water vapor. You cannot see water vapor. It is invisible.

Contrails are line-shaped clouds. They form when gases from jet engines cool and condense high in the air.

Mixing Matter

Another kind of physical change is a mixture. A *mixture* is formed when different kinds of matter are mixed together. Cereal and milk is a mixture. In a mixture, the properties of each kind of matter might change. For example, the cereal might get soggy. However, the milk is still milk and the cereal is still cereal. A mixture can be a combination of solids, liquids, and gases. Vegetable soup is a mixture of liquids and solids. Salad dressing is often a mixture of different liquids. Fog is a mixture of a gas and a liquid. Birdseed is a mixture of several different solids.

This mixture is made of different kinds of fruit.

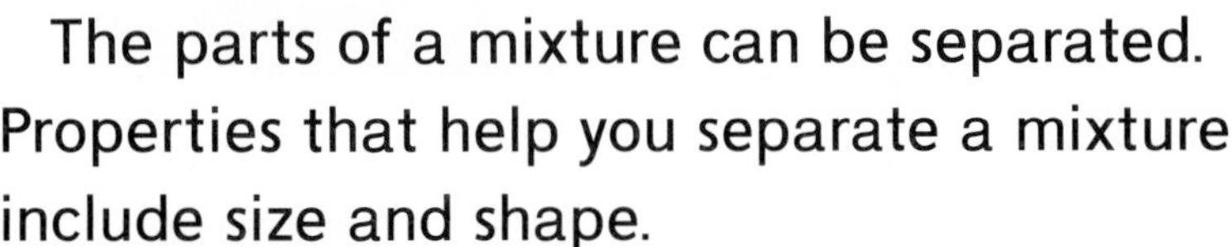

The parts of a mixture can be separated. Properties that help you separate a mixture include size and shape.

Objects that float, such as cranberries, can be separated from objects that sink.

Filters separate mixtures by size.

Magnets separate certain metals from other objects.

Solutions

A *solution* forms when one or more kinds of matter are mixed evenly into another kind of matter. Salt water is a solution. If you add salt to water and stir it, the salt breaks up into very tiny particles. It mixes evenly with the water. You cannot see the salt, but it is still there.

Not all solids form solutions in liquids. If you mix sand and water, the sand will just sink to the bottom. No matter how long your stir, these two materials will not form a solution.

Some solutions contain no liquids at all. Air is a solution of different gases. Brass is a solution of several solids, including the metals copper and zinc.

Just like other mixtures, you can separate the parts of a solution. Think about the solution of salt and water. If you leave that solution, the water evaporates. The salt is left behind.

brass trombone

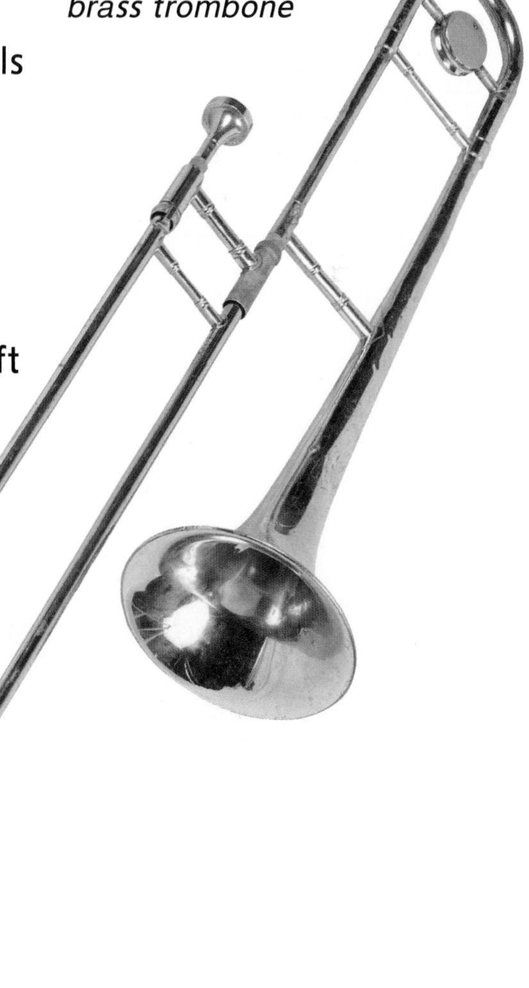

Chemical Changes

During a physical change, the kind of matter stays the same. In a chemical change, the kind of matter changes. Chemical changes cause leaves to change color in the fall. They cause fruit to ripen and meat to spoil. Chemical changes are the reason that cooked food differs from raw food.

Chemical changes are needed for humans to live. Chemical changes in your body keep you warm and give you energy. Chemical changes in plants produce the oxygen that you breathe. They also break down the food you eat into a form your body can use.

Not all chemical changes are useful. They cause food to rot. Objects made of iron rust because of chemical changes. Products you use every day have dates on them. These products might be dangerous to use after those dates because of chemical changes that happen in the products.

A Chemical Change

unripe

ripe

overripe

A chemical change happens when fruit ripens. As bananas ripen, they change color. They also become softer and sweeter.

Skill Builder **Read a Diagram**

Compare the bananas to find differences caused by a chemical change.

Signs of Chemical Change

If one or more of the following things happen, a chemical change might have taken place.

Light and Heat Chemical changes in your body keep you warm. When something burns, light and heat are given off. Chemical changes in a firefly make the insect glow.

Heat and light are given off when wood burns.

Formation of a Gas
Chemical changes in unbaked bread and cake give off gases. These bubbles of gas are trapped when the bread or cake bakes, making little holes in the food. Gases given off when food spoils cause the odor of the food to change.

The holes in this bread are caused by gases given off during a chemical change.

Color Change Color changes can signal a chemical change. A ripe apple is a different color than an unripe apple. Summer leaves on certain trees are a different color than they are in fall. A rusty nail is a different color than a nail that has not rusted.

Chemical change causes green leaves to change to red, yellow, and orange.

Forces and Motion

Position

In this picture, where is the boy in the purple shirt? He is next to the girl in the blue overalls. He is over the girl wearing the pink shirt. When you describe where something is, you describe its position. **Position** is the location of an object.

You can describe something's position by comparing it to the positions of other things. Words such as *over, under, left, right, on top of, beneath,* and *next to* give clues about position. You could say that the pencil sharpener is next to the classroom door. Or you could say that the school cafeteria is to the right of the principal's office. When you describe the position of something, you compare it to objects around it.

How can you describe the position of the girl in the pink shirt?

Purestock/Getty Images

Distance

You can describe something's position by measuring its distance from other objects. **Distance** is the amount of space between two objects or places. Distance can be measured in inches, yards, or miles. In the metric system, distance is measured in millimeters, centimeters, meters, or kilometers. You can use a ruler or a meter stick to measure distances.

Direction

When describing position, you must use both distance and direction. Distance tells how far one object or place is from another. **Direction** tells which way a line points from one object or place to another. The words *north, south, east,* and *west* tell direction. So do words such as *left, right, up, down, forward,* and *backward.* The position of the yellow sponge bear in this picture is to the right of the smaller metal bear.

Make Connections

Read a Diagram
To learn more about measuring distance, look at the sections about SI Units and Customary Units in the Science Guide.

How can you describe the position of the smaller bear, using distance and direction?

1 2 3 4 5 6 7 8 9 10 11 12

Motion

Look at the pictures of the dog below. To the left you can see that the dog is on the ground. Next, you see the dog come completely off the ground. What happened to the dog? It moved. You know that the dog moved because its position changed. While an object is changing position, it is in motion. **Motion** is the process of changing position.

Objects can move in different ways. Look at the chart on the next page. The bowling ball moves in a straight line. Objects can also move round and round, back and forth, or in a zigzag pattern. The skater spins round and round. The snake moves on a zigzag path. It moves forward with short, sharp turns from one side to the other. The pendulum swings back and forth.

How can you tell that the animal has moved?

Brighton Dog Photography/Moment Open/Getty Images

Types of Motion

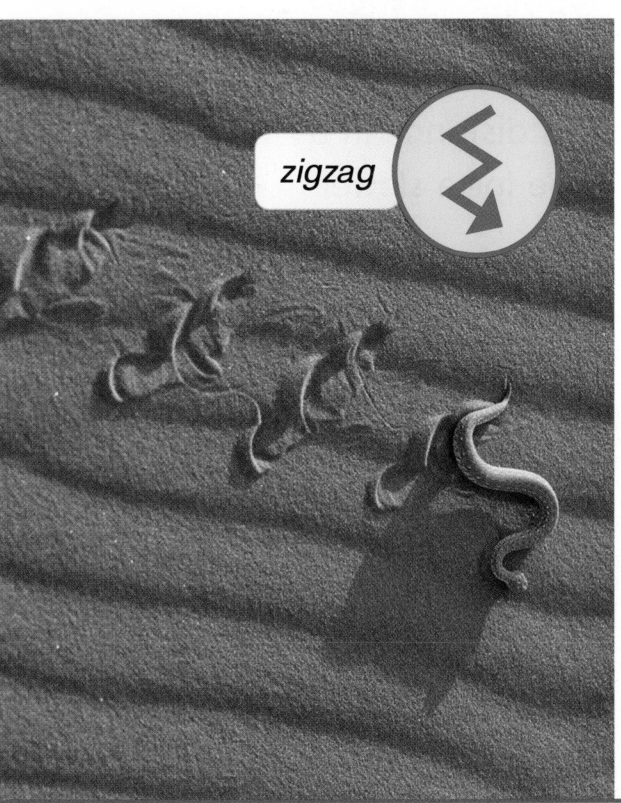

Skill Builder **Read a Chart**

Arrows show direction. Look at the arrows to find out about different types of motion.

Motion - Measuring Motion

Distance You can measure the distance that an object moves. You measure the space between the object's starting position and its new position. Distance measurements are units of length, such as feet, inches, or kilometers. You measure distance the same way you measure length, using a tool such as a meterstick or ruler.

Time Suppose it took you three minutes to walk from your classroom to the playground yesterday. Today it took you five minutes to walk to the playground. You moved the same distance. But your motion today took more time. The time it takes to move a distance is part of the way you describe motion.

Speed Distance and time can be used to find speed. **Speed** is the measure of how fast or slow something moves. An object that is moving fast goes a distance in a short amount of time. It takes a longer time for a slower object to move the same distance.

The swimmers race the same distance. The fastest swimmer reaches the finish line in the shortest time.

Bob Thomas/Getty Images

Direction Direction points out the path from one position to another. Suppose you walk from your classroom to the lunchroom. Later, you walk from your classroom to the music room. Both places are the same distance from your classroom. They both take the same amount of time. Is your motion the same both times? Even if the distance and time are the same, the motion is different. You walk different directions to the two rooms. The direction that something moves is part of the way you describe its motion.

Predicting Motion

Measurements of motion can help you predict future motion. Look at the picture of the girl on the swing. Each time the swing travels back and forth takes the same amount of time. You can predict when she will change direction. You can also predict how much time it will take her to swing back and forth.

Predict how this girl's motion will change next.

Forces

Objects do not move by themselves. A force must be applied to an object to change its motion. A **force** is a push or a pull. When you push on a door handle, you apply a force. The door moves away from you. When you pull on a wagon handle, you apply a force. The wagon moves toward you.

Forces can be large or small. The force that a train engine uses to pull a train is large. The force that your hand uses to lift a feather is very small. It takes larger, stronger forces to move heavy objects than it does to move light objects.

More than one force can push or pull on an object at a time.

Comstock Premium/Alamy

Changes in Motion

Forces can set objects into motion. Forces can also change the motion of objects that are already moving. Forces can affect a moving object's speed and direction. A force can also stop a moving object.

Think about forces that can change the motion of your bicycle. Applying force to the pedals makes your bicycle start moving. Applying force to the brake slows the bike and stops its motion. The more force you use on your pedals, the faster your bike moves. The more force you apply to the brake, the quicker your bike stops.

Forces can also change the direction of motion. You can apply a force to turn the handlebars of your bicycle. This force changes the direction of the handlebars and the front wheel. This movement changes the direction that the whole bicycle moves.

A goalie applies a force to throw the ball. The ball's motion starts.

A player's kick applies force. The ball's motion changes in speed and direction.

Skill Builder **Read a Photo**

Look at the photos to see how forces change the motion of the soccer ball. Captions give information.

Catching the ball stops its motion.

(t) Mike Powell/Stockbyte/Getty Images; (c,b) Ingram Publishing

Types of Forces

Contact Forces

There are many types of forces. The forces you are probably most familiar with are contact forces. *Contact forces* happen between objects that touch. Think about a baseball game. The pitcher must touch the ball to throw it to home plate. A bat must touch the ball to change its direction. A player's glove must touch the ball to catch it.

When the bat hits the ball, it applies a contact force.

Friction

A block slides on the floor. It then slows down and stops. Why does the block slow down and stop? A contact force called friction is acting on the block. **Friction** is a force that occurs when one object rubs against another. Friction pushes against moving objects and causes them to slow down.

Different surfaces produce different amounts of friction. Rough surfaces, such as sandpaper, produce a lot of friction. Smooth surfaces, such as ice, usually produce little friction.

This water slide is smooth, so it has little friction.

Increasing Friction

Sometimes it is helpful to increase friction. People use rough or sticky materials to increase friction. The brakes on a bike use rubber pads to increase friction. When you squeeze the brake handles, the brake pads press against the rim of the wheel. The friction between the pad and the rim causes the bike to stop.

Friction between the brake pad and the bike rim stops the bike.

Reducing Friction

At other times, it is useful to reduce friction. People use slippery things to reduce friction. Oil is often put on moving parts of machines to reduce friction between them. Without oil, friction between the moving parts of the engine will cause the parts to wear out.

If you have been to a hockey game, you might have seen an ice resurfacer used to polish the ice. This machine puts down a thin layer of water to make the ice smoother. This layer of water reduces the friction and helps the players skate faster.

This machine cleans and smoothes the surface of the ice, reducing friction.

Noncontact Forces

Not all forces are contact forces. Some forces can act on an object without touching it. These forces are *noncontact forces.* Magnetism, electricity, and gravity are examples of noncontact forces.

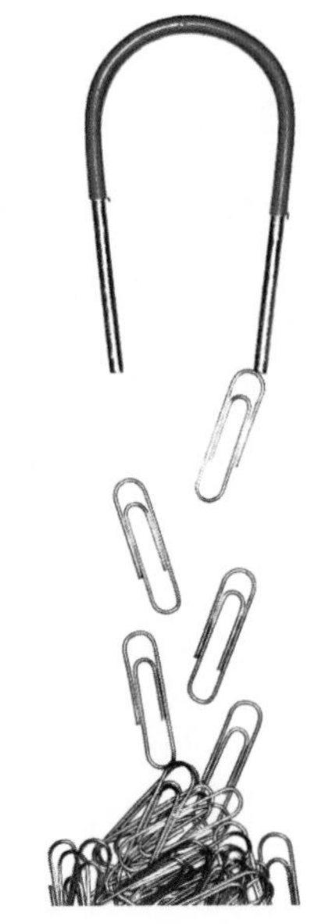

A metal paper clip contains iron. A magnet can pull a metal paper clip without touching it.

Magnetism Have you ever used a magnet? What did you notice? When you bring two magnets near each other, they can **attract**, or pull on, each other. They can also **repel**, or push away from, each other. Magnets do not have to touch to attract or repel each other. The force that causes magnets to attract or repel is called magnetic force. A **magnet** is an object with a magnetic force.

Magnets do not affect only each other. They can also attract objects made of certain metals such as iron. Magnets cannot attract objects made of wood, glass, plastic, or rubber. Magnets can also apply magnetic force through other materials. They can attract or repel objects through solids, liquids, or gases.

Some trains run on magnets that repel them from the track. They produce little friction so they are very fast.

What Could I Be? **Transportation Engineer**

Transportation engineers design vehicles and systems to move lots of people quickly from place to place. They use what they know about forces and friction to build trains and tracks that move as fast as possible with the least friction and the least energy. Learn more about becoming a transportation engineer in the Careers section.

Electricity Objects can have electrical charge. Electrical charge causes electrical force, which is a noncontact force. If two electrical charges that are alike are brought near each other, the charges repel each other. If the two charges are different, they attract each other. Look at the girl in the picture. Each hair on her head has the same electric charge. The like-charged hairs repel each other.

The electrical charges in this girl's hair do not have to touch to affect each other.

Gravity Gravity is another noncontact force. *Gravity* is a pulling force between two objects. Gravity pulls objects, such as you and Earth, together. When you jump up, Earth's gravity pulls you down.

The force of gravity between two objects depends on how much mass the objects have and the distance between the objects. Objects with more mass have a stronger pull. For example, the mass of Earth is huge. Its gravity pulls strongly on all objects. Earth's gravity is what keeps us on the ground. Gravity is also stronger when objects are closer together.

Gravity is pulling these skydivers toward Earth.

Did You Know?

Gravity is one of the forces that keep the planets in orbit around the Sun.

Types of Forces - Balanced Forces

When a batter hits a baseball, he or she applies a force. Other forces, such as gravity, are at work, too. How do all the forces acting on an object affect its motion?

When you put a heavy backpack on your desk, the backpack does not move. Gravity pulls the backpack toward Earth, but your desk is in the way. The desk pushes up on the backpack with a force. The strength of that force is exactly equal to the pull of gravity. The forces are balanced.

Balanced forces are forces that cancel each other out when acting together on an object. The forces are equal in size and opposite in direction. Balanced forces do not cause a change in motion. When an object is sitting still, all of the forces acting on it are balanced.

If the puppies pull in opposite directions with equal force, the shoe does not move. The forces are balanced.

CountryStyle Photography/Vetta/Getty Images

Unbalanced Forces

Suppose you push that heavy backpack across your desk. There is friction between your backpack and the desk. The force of friction is weaker than your push. How can you tell? The backpack moves.

Forces that are not equal to each other are called **unbalanced forces**. If there are two opposite forces acting on an object, then the greater force determines the direction of motion.

Look at the dog sled shown in the photograph. The force of the dogs on the dog sled is greater than the force of friction on the dog sled. So, the dog sled moves. Unbalanced forces can affect an object's direction, speed, or both. The sled now goes in the direction that the dogs run. The faster the dogs run, the greater the speed of the sled.

The dogs place an unbalanced force on the sled.

Types of Forces - Defining Acceleration

You can measure the distance an object moves. You can also measure the time it takes the object to move that distance. From these distance and time measurements, you can determine the speed of the object's motion.

The motion of an object is described by its speed and its direction. Speed and direction can change. Look at the speed skaters in the photograph. As these skaters race around a track, they speed up and slow down. They turn to their left at both ends of the oval track. Any change of their speed or direction is called **acceleration**.

Think about the gas pedal in a car. Another name for the gas pedal is the accelerator. The pedal is also known by this term because it causes the car to accelerate, or change speed. Pressing on the pedal makes the car move faster. Releasing the pedal lets the car slow down.

The skaters accelerate when they change speed and direction around the track.

36clicks/iStock/360/Getty Images

Forces Affect Acceleration

To jump, you push against the ground. To jump higher, you push harder. If you swim, you know that to move faster you must push harder against the water.

The size of a force affects an object's acceleration. A greater force gives more acceleration. The mass of the object matters too. If you apply the same force, an object with more mass will accelerate more slowly.

In the first drawing below, one person pulls the load. The load accelerates. In the second drawing, two people pull the same load. Now it accelerates twice as fast. Why? Two people apply twice the force to the load.

What happens in the third drawing? One person pulls a load that is twice as large. She uses the same amount of force as in the first drawing. The load accelerates half as fast.

Skill Builder

Read a Diagram

Look at the green arrows to determine which load is accelerating fastest.

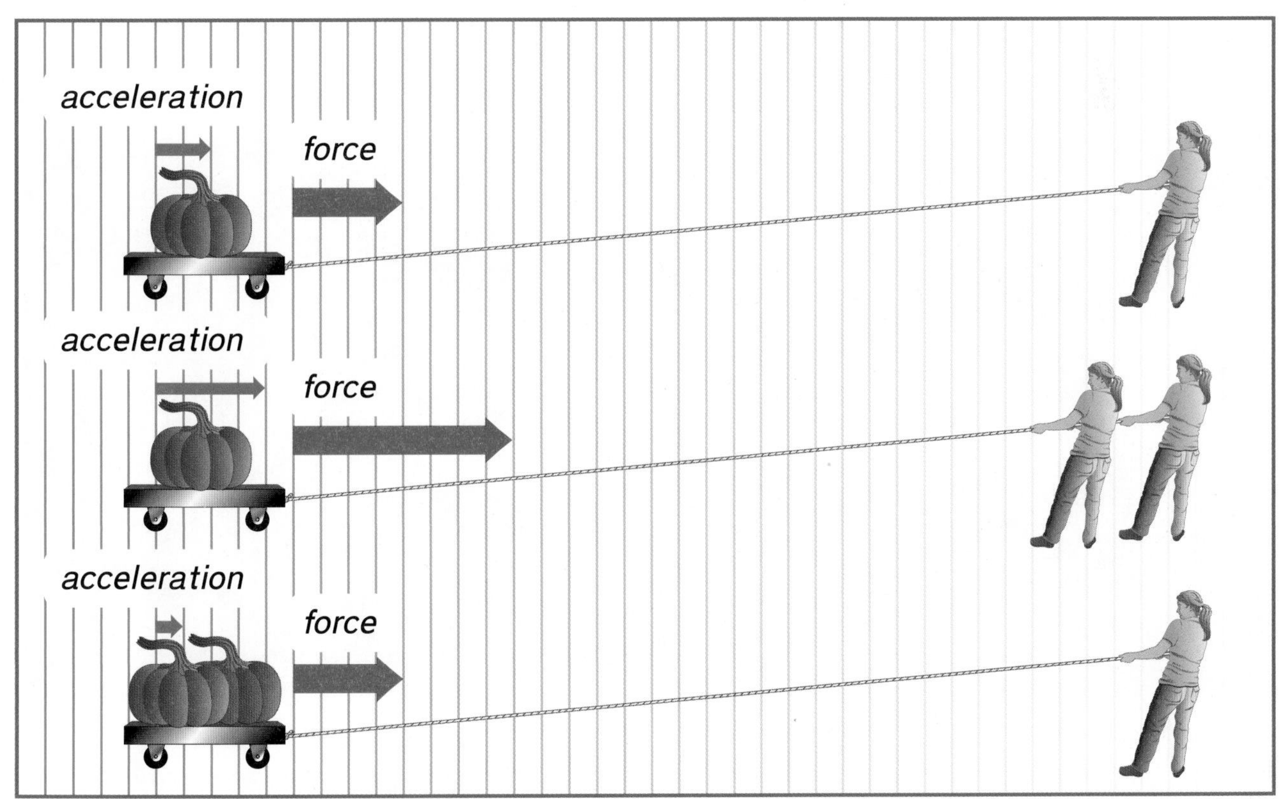

Simple Machines

Work and Machines

In science, the word *work* has a special meaning. **Work** is done when a force moves an object or changes an object's motion. For example, picking up a book is work. A force changes the book's motion. Pushing on a wall is not work. No matter how hard you push, the wall does not move.

A *machine* is a tool that makes work easier. Machines do not change the amount of work done. They just change the way you do the work. For example, it is easier to move a heavy rock by rolling it in a wheelbarrow than it is to lift and carry the rock with your hands.

Some machines help you do work by changing the amount of force you must use. Other machines change the direction in which you push or pull.

Fact Checker

Not all machines are made of many parts and have a motor. Some are simple tools made of just one part.

A backhoe makes moving dirt easier.

A bottle opener is a machine that makes it easier to pry bottle caps off bottles.

Types of Simple Machines

Simple machines are machines with few or no moving parts. There are six types of simple machines. They are the lever, the pulley, the wheel and axle, the inclined plane, the screw, and the wedge.

lever

pulley

wheel and axle

inclined plane

screw

wedge

Simple Machines - Lever

A bottle opener and a seesaw are both levers. A *lever* is a straight bar that moves on a fixed point. The fixed point is the *fulcrum*.

A lever can be used to lift something. The object lifted is called the **load**. In the diagram below, the girl is the load. When the boy presses down on one end of the lever, the load on the other end is lifted. The closer the fulcrum is to the load, the less force you need to lift the load. A crowbar is an example of a lever.

Levers can make it easier for people to lift or pry objects. They can change how much force you need to move something. They can also change the direction of the force you use. Pressing down on a lever lifts up the load.

Skill Builder

Read a Diagram

Compare the arrows. They show how this lever changed the direction of a force.

How a Lever Works

Pulley

A simple machine that uses a rope and a wheel to lift an object is a *pulley.* The rope is looped around the wheel and tied to the object you want to lift. When you pull down on one end of the rope, the other end rises up.

Like some levers, some pulleys change the directions of a force. The wheel of a pulley acts like the fulcrum of a lever.

A pulley makes work easier. Lifting a heavy object directly can be very difficult. Using a pulley lets you use your body weight to help you pull the object. If the wheel is placed above you, the pulley also lets you lift an object higher than you can reach.

A pulley makes it easier to lift this bucket.

Cranes use pulleys to lift heavy loads.

(l) Jose A. Bernat Bacete/Moment Open/Getty Images; (r) Glow Images

Simple Machines - Wheel and Axle

Another special kind of lever is a **wheel and axle**. It is made up of a wheel that moves around a post. The post is called an axle. Doorknobs and Ferris wheels are wheels and axles. A wheel and axle can make work easier to do. Turning a doorknob is easy. Turning the thin bar behind the knob, however, is hard. The doorknob is the wheel. The thin bar is the axle. Turning a wheel requires less force than turning an axle.

Did You Know?

A doorknob uses the wheel to turn the axle. In a Ferris wheel, the axle is used to turn the large wheel. One of the tallest Ferris wheels in the world is the High Roller in Las Vegas, Nevada. Over a thousand people can ride this Ferris wheel at the same time. Think about the amount of force it takes to move that wheel!

Turning the wheel makes the axle of the doorknob turn.

The axle makes a smaller movement. The wheel makes a larger movement.

Inclined Plane

A simple machine with a flat, slanted surface is an **inclined plane.** Inclined planes can make work easier to do. They reduce the force you need to move an object to a higher level. Think about moving a heavy box onto a truck. You could not lift it off the ground to put it in the truck. You could slide it up an inclined plane instead. Sliding a box up an inclined plane requires less force than lifting the box straight up.

Word Study

The word *incline* means "to lean." The word *plane* means a flat surface. So an inclined plane is a flat surface that is leaning.

Screw

An inclined plane wrapped into a spiral is a *screw.* It takes less force to turn a screw than to pound a nail. A screw changes a turning force into a force that pulls the screw into a material.

It takes less force to push a box up a ramp than to lift it straight up.

When you turn a light bulb, the screw end pulls the bulb into the socket.

Simple Machines - Wedge

If you put two inclined planes back to back, you get a wedge. A *wedge* is a simple machine that pushes objects apart. It changes the direction of a force. For example, a wedge changes a downward force to a sideways force.

The head of an axe is a wedge. The axe changes the downward force of the swing into a sideways force. The sideways force pushes, or splits, the wood apart. Most cutting tools, such as knives, are wedges. As you press a knife into food, the knife pushes the food apart into two pieces.

A downward force on the ax pushes the wood left and right.

Pushing down on the knife causes it to push the tomato apart into slices.

Machines Working Together

Most of the tools you use every day are compound machines. A **compound machine** is two or more simple machines put together.

A pair of scissors is a compound machine. Two wedges and two levers make an excellent cutting tool. The point where they are joined is the fulcrum. When you push the handles together, the edges cut through material.

A can opener is also a compound machine. It contains a wedge, a lever, and a wheel and axle. Together they act as one machine.

Skill Builder

Read a Diagram

Notice the labels on each compound machine. Each line points to a simple machine. Think about how these simple machines work together.

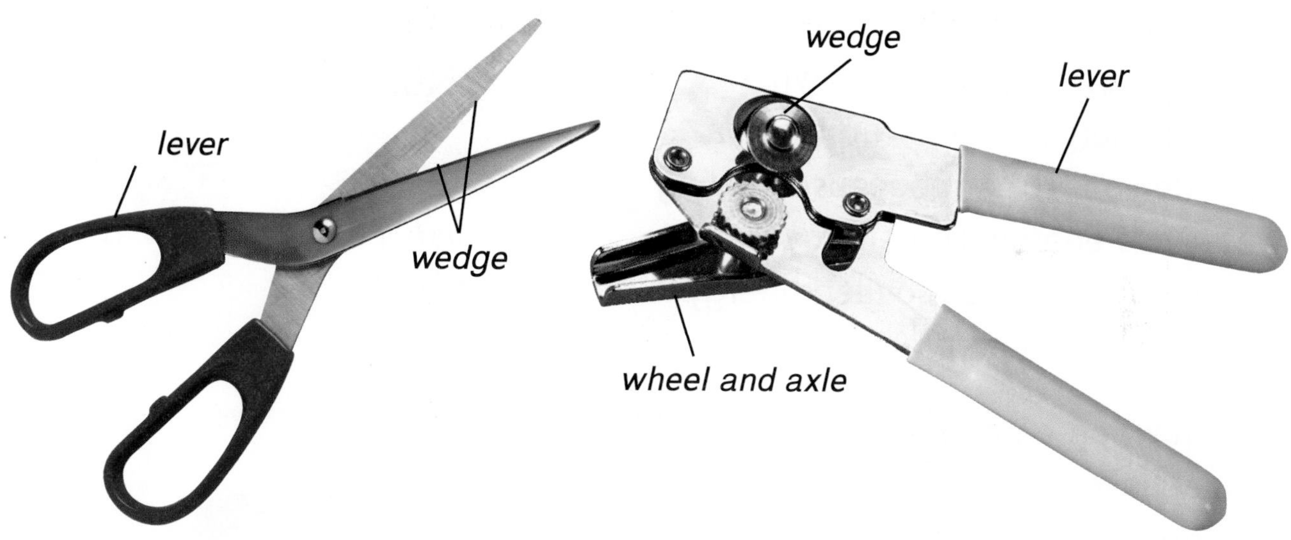

What Could I Be? **Machinist**

Machines make work easier, but who makes machines? Machinists use tools to shape materials. They make the parts that work together to form complex machines. Learn more about becoming a machinist in the Careers section.

Energy

Work cannot be done without energy. **Energy** is the ability to move something or cause a change in matter. An object has energy because of its motion. Condition or position can give objects energy too.

Kinetic and Potential Energy

When matter is in motion, it has energy. A baseball thrown from a pitcher to a catcher has energy. A rollercoaster speeding down a hill has energy. Water can move a boat down a stream because moving water has energy. The energy that moving matter has is called energy of motion, or *kinetic energy*.

When matter is in motion it has kinetic energy.

Even when an object is not moving, it may have the ability to move. *Potential energy* is stored energy that is ready to be used. A diver standing high on a diving board has potential energy. The gasoline that makes a car move has potential energy. The food that fuels your body also has potential energy too.

Potential energy is stored so that the runners can make a quick start.

Image Source

The type of energy objects have is constantly changing from potential to kinetic. For example, when a child climbs the steps to the top of a slide, she has kinetic energy. She is moving. When she sits at the top of the slide, her position gives her potential energy. As she goes down the slide, her energy changes again from potential to kinetic.

Winding the spring gives the toy potential energy. As the spring unwinds, the potential energy changes to kinetic energy.

Winding the spring of a toy changes energy. As the tab is turned, the spring is wound tighter and tighter. The condition of the spring gives the toy potential energy. When the spring is released, it unwinds. The toy moves. Potential energy changes to kinetic energy.

Energy changes as motion and position change.

Moving Energy

Energy can move from one object to another. When you roll a bowling ball, you transfer energy. The energy moves from your body to the bowling ball. When the ball hits the pins, it transfers energy to the pins. The energy causes the pins to move. In soccer, a player transfers energy to the ball by kicking it. Pushing on the pedals of a bicycle transfers energy to the bike. The bike moves.

Energy from the foot makes the ball move.

What Could I Be? Wind Power Worker

Generating power is all about moving energy. Wind power workers design and test blades of the turbines to catch the energy of wind. They use computers to carefully make the blades. They assemble all the parts together and install them in just the right places. To learn more about jobs in the wind-power industry, turn to the Careers section.

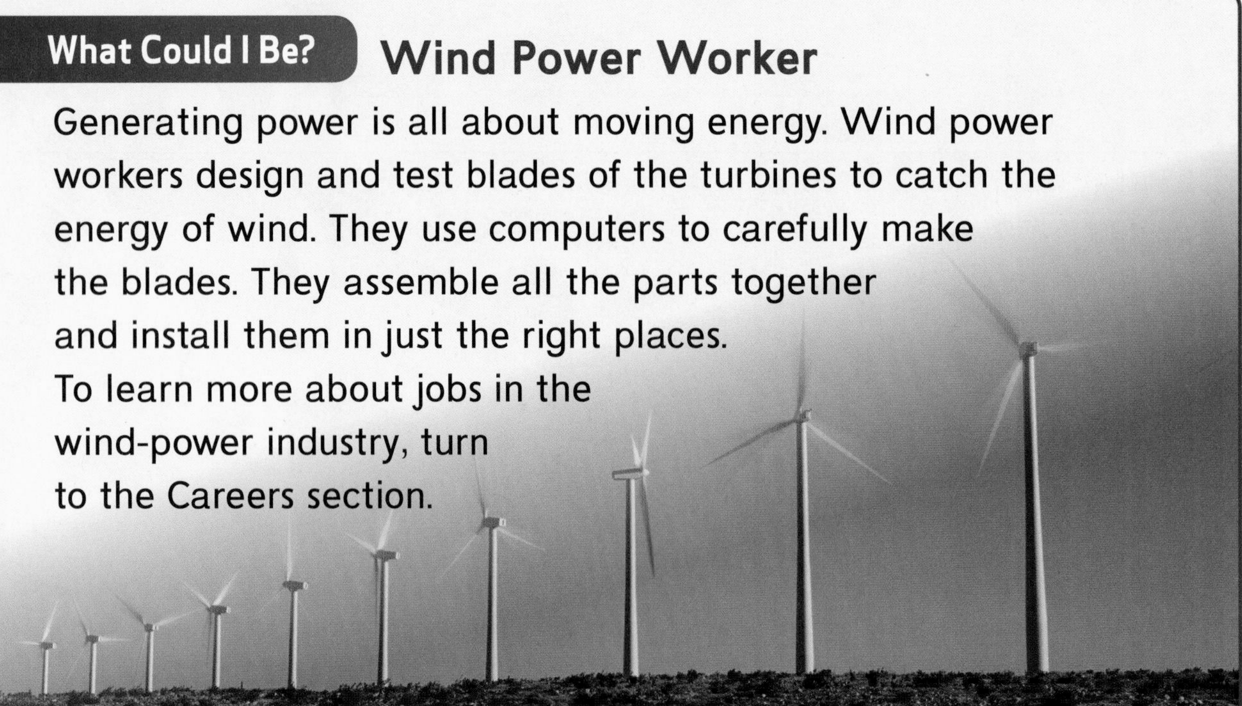

Forms of Energy

Although all energy can be described as either energy of motion or stored energy, there are many different forms of energy.

Form of Energy	Example
Heat Energy All matter is made of small moving particles. The energy of motion of those particles is heat energy. Heat energy moves from warmer objects to cooler objects.	
Electrical Energy Electrical energy is the energy of moving charged particles. Lightning is the movement of charged particles in nature. Electrical appliances use electrical energy that moves through wires.	
Sound Energy Sound is a form of energy that you can hear. Sound moves in waves through matter such as air or water.	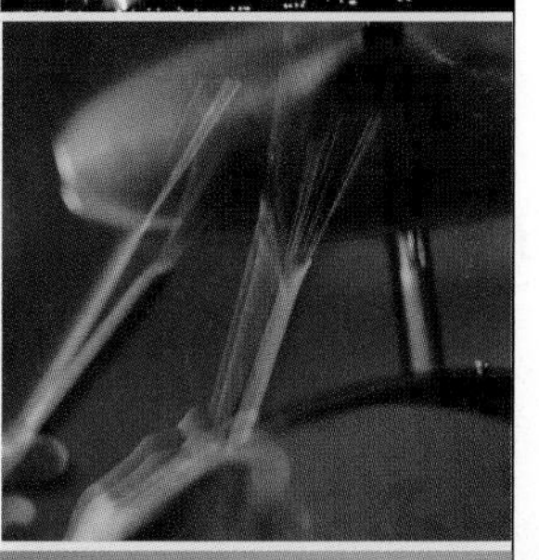
Light Energy Light is a form of energy that you can see. Light energy from the Sun moves in waves through space. Light energy also comes from electric lamps and a blazing campfire.	
Chemical Energy Chemical energy is stored in matter. Fuels such as gasoline, coal, and oil have stored energy. So do matches, wood, and food.	

Heat Energy

Sitting next to a campfire can be very hot. Temperature is a measure of how hot or cold something is. It feels hot near the fire because burning wood gives off heat energy. *Heat* is the flow of energy between objects. Heat can travel through solids, liquids, and gases. It can even travel through space.

No matter what it travels through, heat always flows from a warmer object to a cooler object. If you put your hands around a cup of cocoa, your hands feel warm. Heat energy moves from the hot cup to your cool hands. But suppose you put your hands around a glass of ice water. Your hands are warmer than the glass. Heat energy flows from your warm hands to the glass. Your hands feel cold because they lose heat energy.

The Sun's energy warms Earth's air, land, and water.

Design Pics/Darren Greenwood

Sources of Heat Energy

The Sun is Earth's main source of heat energy. A *source* is where something comes from. The Sun's heat energy warms the air, land, and water. Without the Sun's heat energy, it would be too cold on Earth for most living things to survive.

Make Connections

Jump to the Earth Science section to learn about the Sun.

Fires, appliances, and stoves are some other sources of heat energy. Fires use chemical changes to produce heat energy. Appliances and some stoves change electricity to heat energy.

Rubbing two objects together can also produce heat energy. You experience this when your hands get warm when you rub them together. Machines also produce heat energy when their parts rub together.

Some objects heat up faster than others. When sunlight shines on land and water, the land heats up faster than the water does. That is why the sand at the beach feels hotter than the water.

Heat Affects Matter

All matter is made of tiny particles. These particles are always moving. The energy that makes them move is *thermal energy*. Heating matter increases how much thermal energy the particles have. A hot object's particles move quickly. A cold object's particles move slowly.

Thermal energy is what makes objects feel hot or cold. When you measure an object's average temperature, you are measuring its thermal energy. For the same amount of material, the object with more thermal energy will have a higher temperature.

Matter gains energy when it is heated. The energy can cause its temperature to rise. Cooling matter causes it to lose energy. Its temperature drops. At certain temperatures, matter will change state. The temperature at which different types of matter changes state is different and depends on the type of matter.

Skill Builder

Read a Diagram
Look carefully at the particles. The trails behind some of the particles show that those particles are moving quickly.

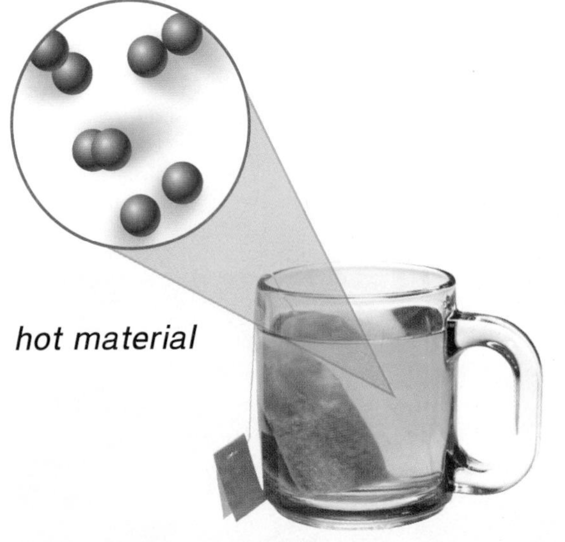

hot material

cold material

John A. Rizzo/Getty Images

Expanding and Contracting

When heat flows into an object, the object gains thermal energy. Its temperature increases. The particles in the object move faster and farther apart. The increase in energy causes most objects to get bigger, or *expand*.

When heat flows away from an object, the object loses thermal energy. Its temperature decreases. Its particles move more slowly. The decrease in energy causes most objects to get smaller, or *contract*.

A thermometer is a tool that measures temperature. Some thermometers are made up of a clear tube filled with a liquid. When the temperature of the liquid increases, the liquid expands. It fills more of the tube. When the temperature of the liquid decreases, the liquid contracts. It fills less of the tube. The change in the level of the liquid shows the temperature.

Skill Builder

Interpret a Photo

To read the thermometer, line up the top of the red liquid with the white markings on the thermometer.

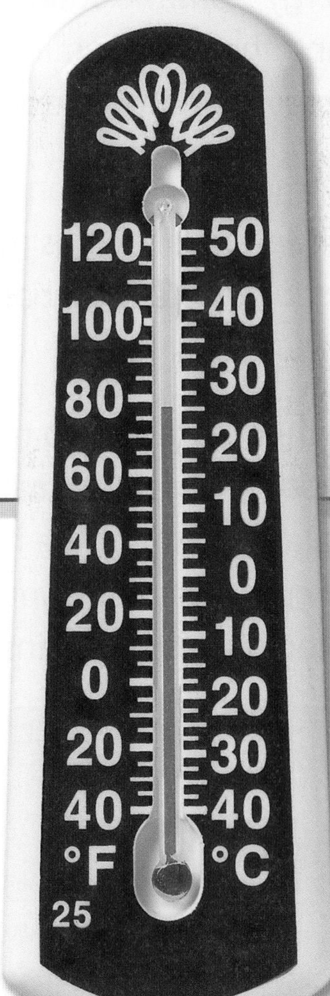

Fact Checker

Water is different from most types of matter. When water freezes, it expands. When ice melts, it contracts. Expansion and contraction of water causes roads and sidewalks to crack as temperatures change.

(r) Ken Cavanagh/McGraw-Hill Education; (l) juan angel de corral/Moment Open/Getty Images

How Heat Moves

Heat energy moves more easily through some materials than through others. Heat energy also moves differently through different types of matter.

Conduction

Touching a metal spoon left in a bowl of hot soup provides evidence that heat energy moves. Heat energy travels by *conduction* through materials that are touching. The heat energy from the soup moves into the spoon. Then it moves along the spoon until the whole spoon is hot. The heat energy then travels from the hot spoon to your hand.

Some materials conduct heat energy better than others do. Materials such as metals are good conductors. A *conductor* is a material that heat energy moves through easily.

Other materials do not conduct heat energy well at all. These materials are called *insulators.* A foam cup is an insulator. It keeps the heat energy of a hot beverage inside the cup, away from your cooler hand. Wood, glass, air, and water are also good insulators.

The pan is a good conductor of heat. It allows heat to move from the oven to the food being cooked. The oven mitts are good insulators. They protect the hands of the person touching the hot pan.

YinYang/iStock/Getty Images Plus/Getty Images

Convection

The movement of heat energy in liquids and gases is called *convection.* Warm liquid or gas is forced up by cooler liquid or gas. For example, when water boils in a pot, the water near the heat source gets hot first and is forced up by the cooler water. Then the cool water sinks, gets heated, and is forced up. This movement happens over and over in a circular current, or flow.

Radiation

Heat can also move without any matter to carry it. For example, the Sun gives off *radiant energy.* Some of that energy is the light you can see. However, much of the energy is in the form of invisible waves. These waves travel through space. The movement of heat energy in the form of waves is called *radiation.* The sun is not the only source of radiant energy. Any light that shines on you can warm you by radiation.

The heating of water in a pan on a stove shows all types of heat movement. The stove heats the pan through conduction. The stove heats the air around the pan by radiation. The water at the bottom of the pan heats by conduction. The rest of the water heats by convection.

Electricity and Magnetism

Electrical Energy

Matter is made up of very tiny parts, called particles. These particles have either a positive or a negative charge. Electrical energy is the energy of these charged particles.

Electric Charge

You feel a shock when an electrical charge moves from a doorknob to your hand.

Frank Rosenstein/Photodisc/Getty Images

You sometimes feel a shock when you touch a metal doorknob. This shock occurs because of electricity. If the room is dark, you can see a quick spark of light with the shock. A flash of lightning and the glow of a light bulb also occur because of electricity.

The property of matter that causes electricity is an **electrical charge**. You cannot see electrical charge, but you can understand how objects with different charges interact.

There are two types of electrical charges. One is called positive. The other is called negative. The diagram shows how objects with positive or negative charges affect each other.

An object with a positive charge and an object with a negative charge attract.	+ → ← −
Objects that both have a positive charge push each other away.	← + + →
Objects that both have a negative charge also push each other way.	← − − →

Static Electricity

All objects are made of charged particles. Most objects have the same numbers of positive particles and negative particles. The charges are balanced. When two objects touch, negative particles can move from one object to the other. Negative particles may build up on one object. That object has a negative charge. A buildup of electrical charge is called **static electricity**.

Lightning occurs because of a buildup of electrical charge in clouds.

A *discharge* occurs when static electricity moves from one object to another. This movement is why touching a doorknob can sometimes produce a shock. As a person walks across a floor, negative particles move from the floor to the person. These negative particles move to the doorknob when it is touched. A lightning strike is an example of a very large discharge.

Make Connections

Jump to the Earth Science section to learn more about lightning.

When you rub a balloon on a sweater, negative (-) particles move from the sweater to the balloon. Then if you hold the balloon near a wall, the negative charge attracts the positive (+) particles on the wall. This attraction causes the balloon to stick to the wall.

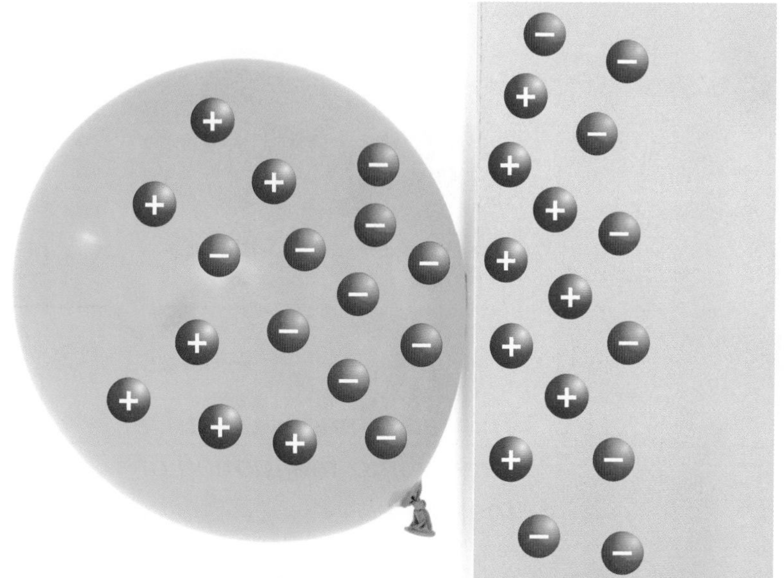

Electrical Energy - Electric Current

Charged particles can build up on objects. They can also be made to flow. A flow of charged particles is called an *electric current*. You use electric current every day. Electric current provides the energy you need to power lights, radios, computers, hair dryers, and many other products. We use energy from electric current to produce heat, light, sound, and motion.

Electrical energy changes to heat energy inside this toaster.

This lamp changes electrical energy into light.

These headphones change electrical energy into sound.

This fan changes electrical energy into the motion of the blades.

Circuits

Electric current needs a path through which to flow. A *circuit* is a path that is made of parts that work together to allow current to flow. Simple circuits have several parts. A battery may be the circuit's source of power. Wires connect the different parts of the circuit.

To keep an electric current moving, a circuit cannot have any breaks. A *closed circuit* is a complete, unbroken circuit. An *open circuit* is a circuit with breaks.

A *switch* allows you to control the flow of current. When the switch is in the "on" position, there is no gap in the path. The circuit is closed, and the current can flow. Turn the switch off, and there is a gap in the path. The circuit is open, and current does not flow.

What Could I Be?

Electrical Engineer

Electrical engineers design and build circuits or electrical devices. Examples of systems an electrical engineer might work on include computers, robots, medical equipment, and phones. Learn more about electrical engineering in the Careers section.

open circuit

When the switch is open, electric current does not flow. The bulb does not light.

closed circuit

When the switch is closed, electric current flows. The bulb lights.

Electrical Energy - Conductors and Insulators

The electric current in a home flows through wires. These wires are usually made of copper and are wrapped in plastic. Copper is used because current can easily flow through it. Copper is a conductor. A conductor is a material that allows current to flow easily through it. Most metals are conductors.

The copper wires in your home are coated in plastic. The plastic coating on wires does not let current flow through it. Plastic is an insulator. An insulator is a material that does not allow current to flow easily through it. The coating on wires in your home protects you from getting shocked. The table shows examples of good conductors and insulators.

Examples of Good Conductors	Examples of Good Insulators
• copper • aluminum • iron • silver	• glass • plastic • rubber • wood

Copper is used to make wires because it is a good conductor.

Electrical plugs are covered in plastic because it is a good insulator.

Magnets

A magnet is made of material that can attract iron and certain other metals. The ability of an object to push or pull on another object that has the magnetic property is called magnetism. Magnets can attract and repel each other with magnetic forces. Objects attract if there is a force that pulls them toward each other. Objects repel if there is a force that pushes them apart.

Magnets can be made in different shapes and sizes. Some have a bar shape. Others have the shape of a horseshoe or a ring. Many electronic devices have magnets built into them that you cannot see.

Magnetic force causes magnets to attract objects made of certain types of metal.

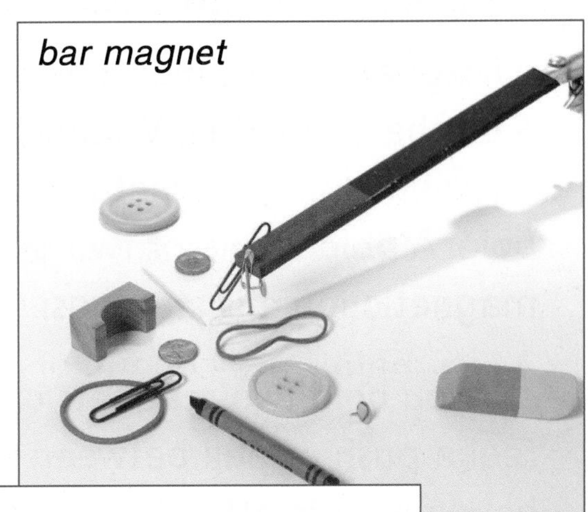

bar magnet

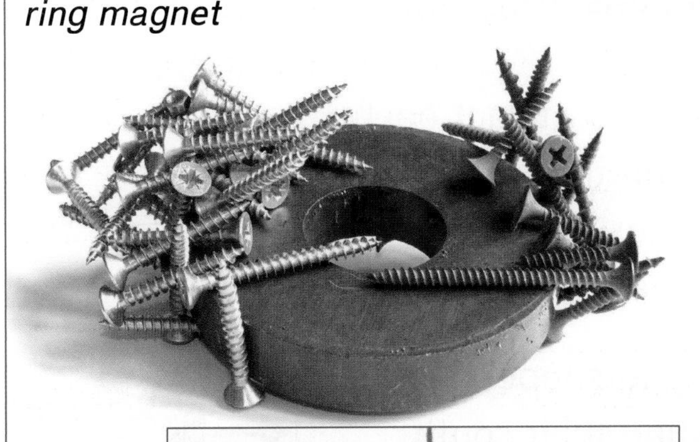

ring magnet

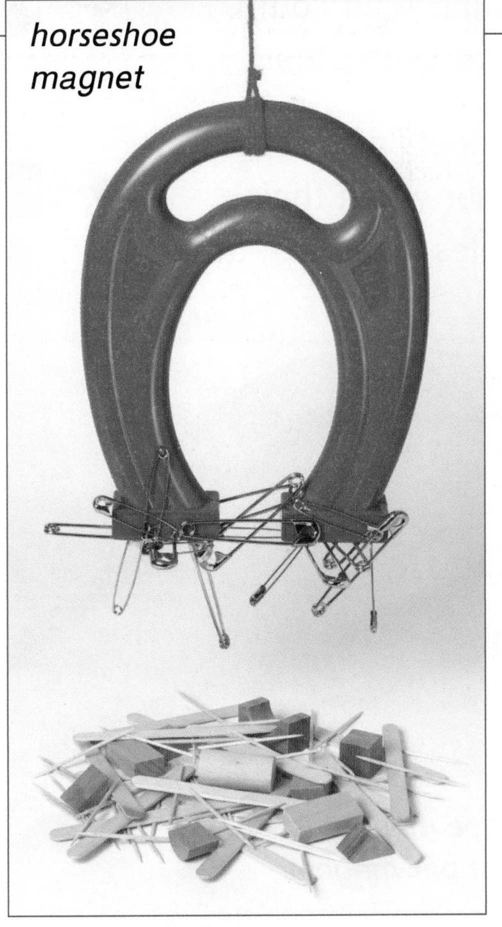

horseshoe magnet

Magnets - Magnetic Poles

Magnets sometimes have *N* painted on one end and *S* on the other. The *N* stands for *north,* and the *S* stands for *south.* Each magnet has a north pole and a south pole. A **pole** is one of two ends of a magnet where the magnetic force is strongest.

If you hold two magnets close to each other, you can feel a push or pull between them. The diagram shows how magnets attract or repel each other.

Two magnets attract each other when the south pole of one faces the north pole of the other.

Two magnets repel each other when their south poles face each other.

Two magnets also repel each other when their north poles face each other.

A horseshoe magnet is like a bent bar magnet.

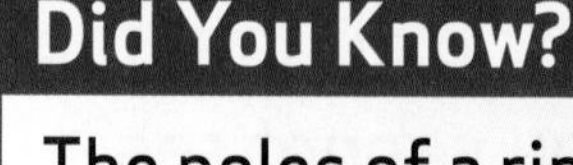

Did You Know?

The poles of a ring magnet are on its flat sides. One side is the north pole. The other side is the south pole.

Ken Cavanagh/McGraw-Hill Education

Magnetic Field

If you want to throw a ball, your hand has to touch the ball. A magnet can push or pull an object without touching it. It does have to be close enough to the object to be in their magnetic field.

A **magnetic field** is the area around a magnet where its force can attract or repel. You cannot see a magnetic field, but you can feel where it is. If you bring two magnets close, you can feel them push or pull each other. Even when the magnets do not touch, their magnetic fields interact. If you move the magnets apart, you do not feel the push or pull any longer. The magnetic fields are no longer meeting.

Bits of iron were sprinkled around this magnet. The bits of iron show the magnetic field.

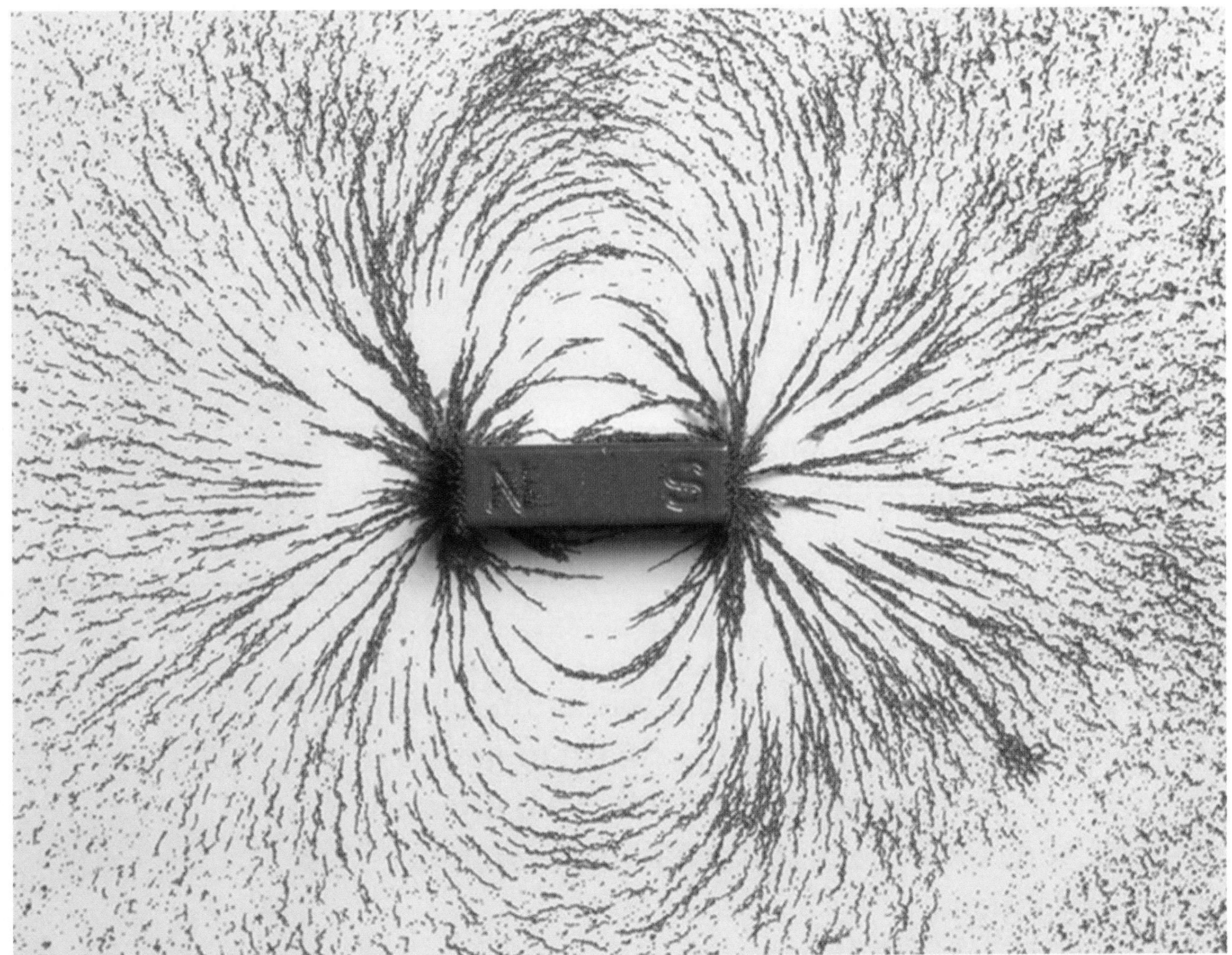

Lars Nikki/McGraw-Hill Education

Magnets - Earth's Magnetic Field

Earth is a giant magnet. Iron deep inside Earth creates a huge magnetic field around the planet. Just like a bar magnet, Earth has two magnetic poles.

Earth's magnetic poles are not at exactly the same locations as Earth's geographic poles. Earth's geographic poles, the North Pole and the South Pole, are located at the north and south ends of Earth's axis. The magnetic poles are in slightly different locations.

Earth is surrounded by a magnetic field.

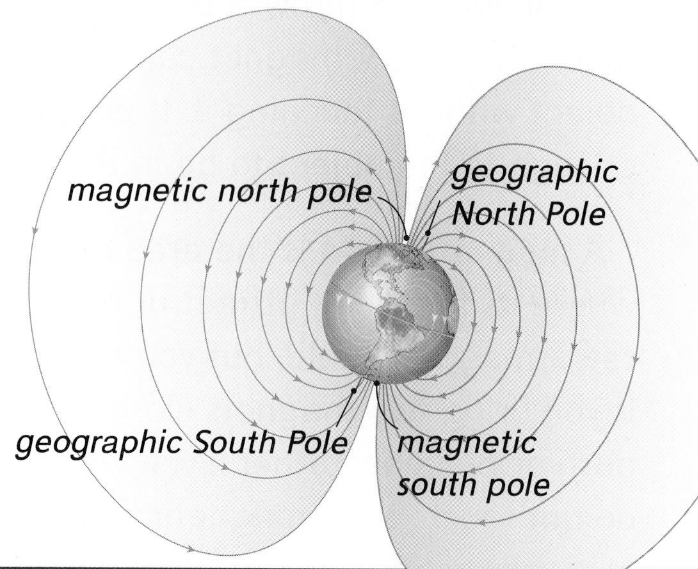

Fact Checker

We call the magnetic pole near Earth's geographic North Pole the "magnetic north pole." It attracts the north pole of magnets though. That fact means that it actually behaves like the *south pole* of a magnet because opposites attract.

Magnetic Compass

A *compass* is a tool that shows direction. It uses Earth's magnetic field. The needle of the compass is a magnet that can move around. The tip is the north pole of the magnet. It points generally toward Earth's geographic north.

A compass uses magnetic force to show direction.

What Could I Be? Navigator

A navigator is the person aboard a ship or an aircraft who is responsible for plotting the course to get from one place to another. Earth's magnetic field affects the tools and calculations of navigators. Learn more about navigation in the Careers section.

Electromagnets

A magnetic field forms around a wire if a current flows in the wire. If you wind the wire into a coil, the field is stronger. When a current flows in the wire, the coil becomes a magnet. The magnet is stronger if you place a metal bar inside the coil. An *electromagnet* is a coil of wire around a metal bar such as iron. A battery at the ends of the wire makes a current flow in the wire.

A simple electromagnet is a coil of wire around an iron bar. The battery makes a current flow in the wire.

You can turn an electromagnet on and off with a switch. The switch makes electromagnets useful in many electric devices.

An electromagnet attracts certain metal, just as a bar magnet does.

Sound and Light

Sound Energy

To *vibrate* is to move back and forth quickly. *Sound* is a form of energy that comes from objects that vibrate. Place your hand gently against your throat, and make a humming sound with your voice. You can feel the vibration and hear the sound.

When you pluck the strings of a guitar, the strings vibrate to produce sound.

How Sound Travels

When you drop a stone into water, waves move out in all directions. Sound travels in waves too. When you pluck a guitar string, it vibrates. The vibrating string bumps into particles of air. Those particles bump into other air particles. This movement produces a sound wave that moves through the air. You hear the sound when the wave moves the particles of air in your ear.

Echoes If you throw a ball toward a hard surface, the ball bounces back. Sound waves also bounce back when they hit a hard surface. An *echo* is a sound wave that bounces back toward you. You may hear an echo when you talk in an empty room.

Did You Know?

A bat finds its prey by sending out sound waves and listening for the echo.

Sound Travels Through Matter When you hear a car motoring down the street, the sound waves are traveling through the air. Air is a gas. Sound waves can travel through gases, liquids, and solids. Some sea animals communicate by making sounds underwater. In fact, whales can communicate through miles of water! You hear a knock on the door because sound travels through the solid door and then through the air. Sound waves travel through all types of matter, but sound does not travel through completely empty space.

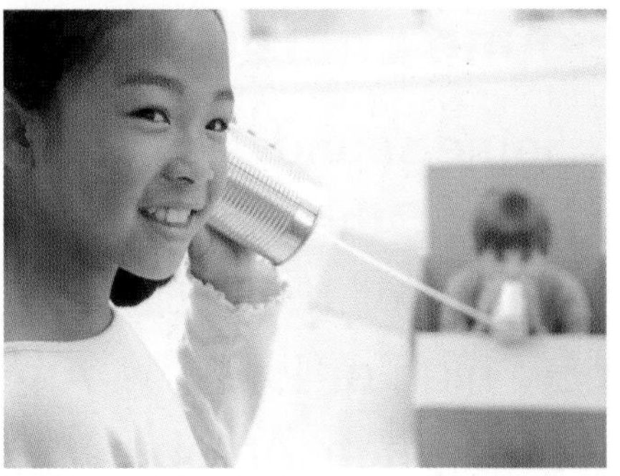

These students can hear sound because it travels through the solid string. The cans vibrate to amplify the sound. The cans also direct the sounds toward the students' ears.

Speed of Sound Sound does not travel at the same speed through all materials. Sound travels slowest through a gas. It travels faster through a liquid, but it travels fastest through a solid. The varying speeds of sound waves make different sounds when you hear them through different materials.

Orca whales use sounds to communicate.

(b) Rick Rhay/Getty Images; (t) Fancy Collection/SuperStock

Sound Energy - High and Low Sounds

Some sounds are high, such as the squeaking of a mouse. Other sounds are low, such as the rumble of thunder or the croaking of a bullfrog. A sound's *pitch* is how high or low it is. An object that vibrates quickly has a high pitch. An object that vibrates slowly has a low pitch.

Your ears cannot hear all sounds. Some sounds have a pitch that is too high for you to hear. Some sounds have a pitch that is too low for you to hear. Many animals, however, can hear sounds that people cannot hear.

Did You Know?

Wind instruments produce sound when air is forced through the pathways inside of them, causing the air to vibrate.

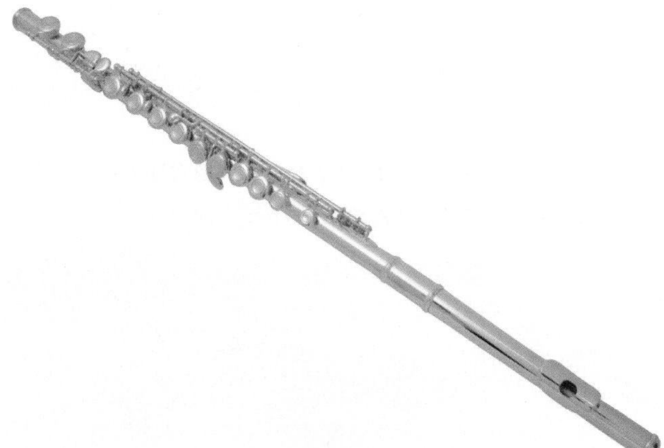

A flute makes sounds with a high pitch.

A tuba makes sounds with a low pitch.

A dog can hear sounds with a pitch that is too high for humans to hear.

An elephant can hear sounds with a pitch that is too low for people to hear.

Changing the Pitch of Sound

Different factors affect the pitch of sound. A vibrating object's length can affect pitch. Look at the marimba in the picture. When hit, the shorter keys vibrate faster than the longer ones. The shortest key vibrates fastest and produces the highest pitch.

The thickness of an object can also affect its vibration and pitch. A guitar has both thin strings and thick strings. Thin strings vibrate faster than thick strings. Thin strings have higher pitch. A guitar player can change the pitch of a string by tightening or loosening it. Pitch gets higher when a string is tighter. It gets lower when a string is looser.

When you speak, air rushes past muscles in your throat called vocal cords. You change the pitch of your voice by making the muscles tighter or looser.

A marimba can make sounds with high to low pitch.

This singer can produce high and low notes with her voice.

Sound Energy - Loud and Soft Sounds

Sounds can have different volumes. *Volume* describes how loud a sound is. A plane flying overhead is louder than a bird's song. A plane has a greater, or higher, volume.

Objects that vibrate with a lot of energy make loud sounds. The more energy an object vibrates with, the louder the sound it makes. You notice this fact if you first tap your foot on the floor and then stomp on the floor. You use more energy to stomp your foot than to tap it. Stomping makes a higher-energy vibration, so you hear a louder sound. Tapping makes a lower-energy vibration, so you hear a softer sound.

The volume of a sound does not change its pitch. A piano player, for example, can play a sound with a certain pitch very loud. The player can then play the same pitch so soft that you barely hear it.

Planes have sounds so loud that you can hear them from miles away.

Sounds in an art museum usually have a low volume.

Sound and Distance

You can hear a sound better when you are close to the source of the sound. When you stand near a school bell, for example, the ring will sound very loud. Farther away the ring does not sound so loud. The volume of the bell has not changed, but what you hear has changed. Loudness decreases as you move away from the source of a sound. The sound's energy spreads out as the sound waves move away from the source so the sound is not as loud.

The closer you get to a school bell, the louder it will sound.

Hearing Sound

The diagram shows how sound waves travel through the human ear. Remember, sound waves vibrate air particles. Your outer ear works like a funnel to collect the vibrating air. Next, the sound wave makes your eardrum vibrate. Your eardrum vibrating causes three tiny bones in your inner ear to vibrate. These vibrations pass through the inner ear to nerves. The nerves send a message to your brain, and you hear a sound.

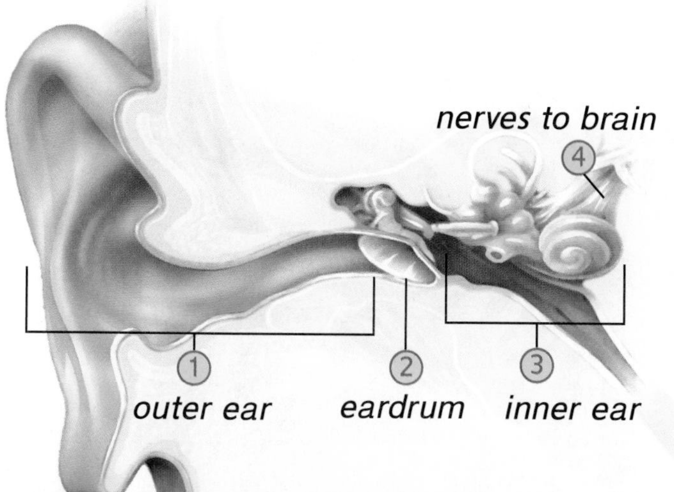

What Could I Be?

Audiologist

People's ears do not always detect sound well enough for them to hear clearly. An audiologist is a medical professional who helps people by diagnosing and treating hearing problems. Learn more about the work of an audiologist in the Careers section.

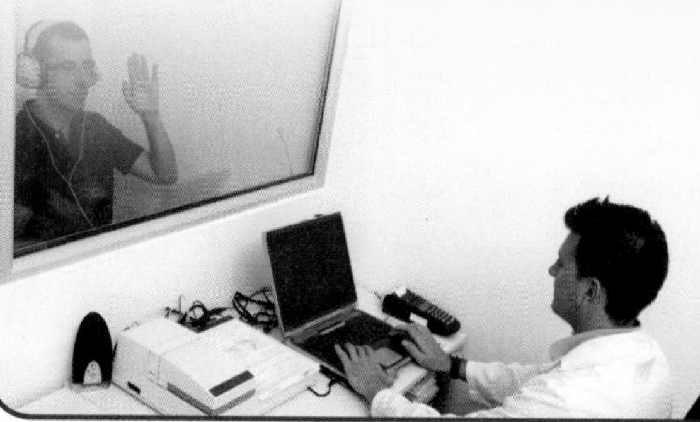

Skill Builder: Read a Diagram

Follow the numbers to see how sound waves travel through the ear.

Light Energy

Light is a form of energy. It allows you to see objects. Light comes from many different sources. The Sun is the source of light that makes Earth's daytime bright. Fires and light bulbs are some other sources of light.

Light moves away from this lighthouse in a straight path.

Light travels away from its source in beams that follow a straight path. When you turn on a flashlight, you can see a straight beam of light. Even light from the Sun travels millions of miles through space in a straight path. Light travels in a straight path until it hits an object.

Absorption

Light can be *absorbed,* or taken in, when it hits an object. Black objects absorb almost all the light that hits them. Remember that light is a form of energy. When a black object absorbs light, this energy heats the object. The energy causes the object to become warmer. White objects absorb almost no light.

The black stripes of this zebra absorb almost all the light that strikes them. The white parts absorb very little light.

Reflection

Most objects do not make their own light. You see those objects when light *reflects,* or bounces off them, and goes into your eyes. The light changes direction and then moves in a new straight path.

Light bounces off objects in the same way that a ball bounces. Suppose you drop a ball straight down onto a hard, flat surface. The ball will bounce straight back up. If, instead, you bounce the ball down at an angle, it will bounce away from you at that same angle. When light hits an object, it bounces off in a different direction. The direction depends on the way the light hit the object.

Did You Know?

The Moon does not make its own light. Moonlight is actually sunlight that reflects off the surface of the Moon toward Earth.

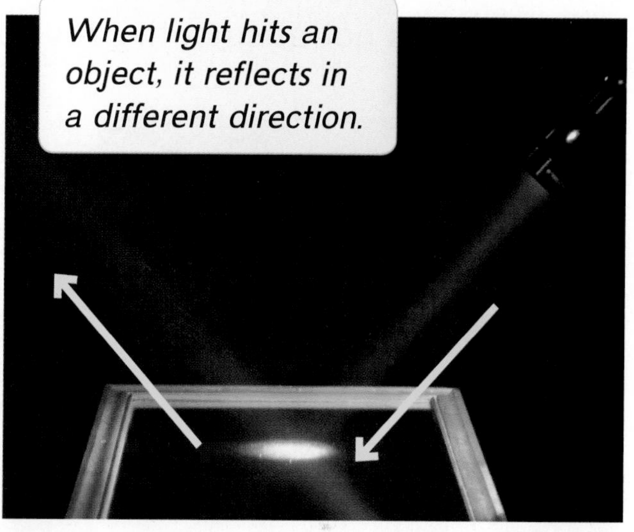

When light hits an object, it reflects in a different direction.

The ball bounces straight back when it hits the flat surface in a straight line. Light bounces the same way off a flat surface.

This ball will bounce away at the same angle as it hits the backboard. Light also bounces away at the same angle as it hits a flat surface.

Light Energy - Surfaces That Reflect Light

A mirror is a very smooth, shiny surface. Mirrors reflect almost all of the light that strikes them. Mirrors do not produce their own light. When you see an image of yourself in a mirror, you are seeing light that bounces off the mirror and into your eyes.

You see your image in a mirror because the smooth and polished surface reflects light evenly. Light bouncing off a surface that is not as smooth does not reflect evenly.

The cat sees its image because the mirror reflects light evenly.

The smooth water shows an image because it reflects light evenly.

No image is reflected when the water is not smooth. Light bounces off the surface in many directions.

(tl) D.Jiang/Moment Open/Getty Images; (bl) Ryan McVay/Getty Images; (r) Jessica Byrne

Blocking Light

Opaque Objects Light cannot travel through many materials. An *opaque* object blocks light from passing through. A brick wall, a piece of cardboard, and your body are opaque. You cannot see through opaque objects.

Shadows Opaque objects can cause shadows to form. A *shadow* is a dark space that forms when light is blocked. You have probably seen your shadow on a sunny day. Your body blocked the sunlight. The shadow that formed had a shape similar to your body.

A shadow's size depends on where the light source is. The closer an object is to a light source, the bigger the shadow. Light coming from above produces a short shadow. As the light source gets closer to the ground, the shadow gets longer.

Make Connections

Read a Diagram

Jump to the section The Moon to learn how shadows cause the phases of the Moon.

This door is opaque. Light cannot travel through it.

A shadow forms on the opposite side of an object from the light source.

Light Energy - Transparent Objects

Light can pass through some objects. *Transparent* objects allow light to pass straight through them. Air, glass, and clear plastic are examples of transparent materials. You can see clearly through these objects because they allow almost all light to pass through them.

Translucent Objects

You see a blurry image when you look through some objects. *Translucent* objects let some light pass through them, but they scatter the light. You cannot see clearly through the objects. Wax paper and frosted glass are examples of translucent materials.

Word Study

Trans- means "through" and *-lucent* means "to shine." So *translucent* means "to shine through."

A transparent glass window allows you to see the fish inside this tank.

You cannot see the image as clearly through this translucent window.

Refraction

Light can refract when it moves from one material to another. To *refract* means "to bend," which means the light changes direction. The amount that it bends depends on the two materials the light is passing through.

Look at the picture of the pencil in the water. The pencil looks as if it is broken, but it is not. You see light from the pencil through the glass. Light bends when it passes from the air through the glass. It also bends when it passes from the glass through the water. But it bends differently through the different materials, so you see the pencil a little differently through them. The difference in the bending of light makes the pencil look broken.

Bending light make this pencil look broken.

A fish in a fish tank is not where it seems to be. Light bounces off the fish into your eyes. The light, however, bends when it passes from the water to the air.

Light Energy - Colors in White Light

Sunlight looks white, but in fact, it is a mixture of many colors. White light, such as sunlight or light from a lamp, is made up of every color of light.

Prisms A prism reveals the colors that make up white light. A *prism* is a piece of glass that refracts light. Prisms separate white light into all the colors that make it white. They do this by bending each color of light by a different amount. Violet light bends the most. Red light bends the least.

When light passes through a prism, it separates into different colors.

Rainbows A rainbow also shows the colors of white light. Water droplets in the sky can act as tiny prisms. When the water droplets refract sunlight, the colors that make up the light spread out. A rainbow forms.

Did You Know?

A rainbow is curved because the droplets of water that form it are curved.

A rainbow can form only when sunlight shines onto tiny droplets of water.

The Color of an Object

Objects have many different colors. A banana is yellow. A leaf is green. The color of an object depends on the light that it absorbs and reflects. You do not see the colors of light that are absorbed. You only see the colors of light that are reflected.

red

The apple is red because only red light reflects off it.

black

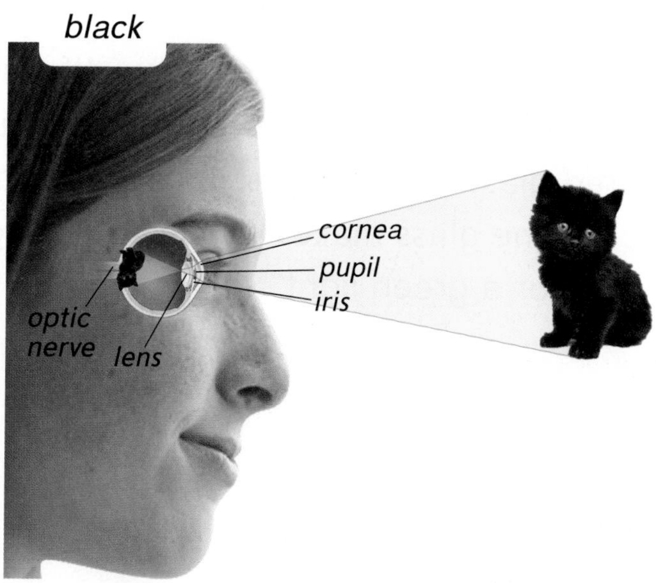

The kitten is black because almost no light reflects off it.

green

The leaf is green because only green light reflects off it.

white

The snowball is white because almost all light reflects off it.

Light Energy - Color Through a Filter

The color of an object can look different if the light moves through a filter. A *filter* is a material that lets only certain colors of light pass through it. A filter may be a piece of colored glass. Red glass, for example, lets only red light through. It blocks the other colors of light.

The three lights in this stoplight are all white. When one of the lights turns on, the light passes through red glass, yellow glass, or green glass. The color of the glass makes you see a red light, a yellow light, or a green light.

Colored glass makes a white light look red, yellow, or green.

The glass on this lamp shade filters out some colors of light and lets other colors pass through.

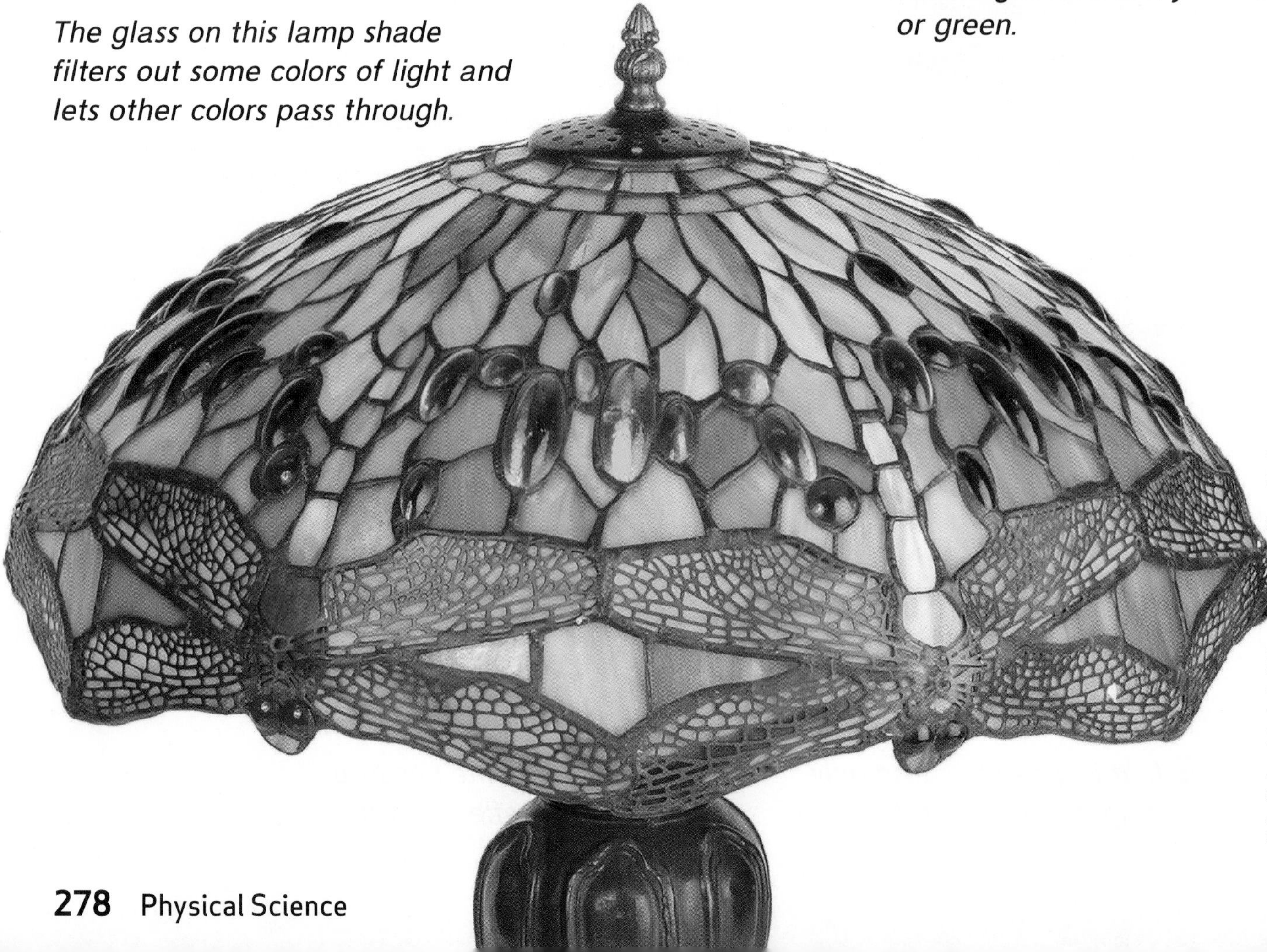

(t) Corbis/age fotostock; (b) Ingram Publishing/SuperStock

How You See

Light allows you to see objects. Some objects, such as the Sun, make their own light. Other objects, such as a book, do not make their own light. You see the object because it reflects light from another source. That reflected light travels into your eyes.

The diagram shows different parts of the eye. Light that reaches your eyes refracts when it goes into the clear *cornea.* Then, light passes through the pupil. The pupil is the black opening in the center of the eye. The pupil controls how much light enters the eye. Next, light travels through the lens. The lens refracts the light so it strikes the back of the eyeball. The optic nerve sends information about the light to the brain. The brain then translates that information to understand it as a picture or image.

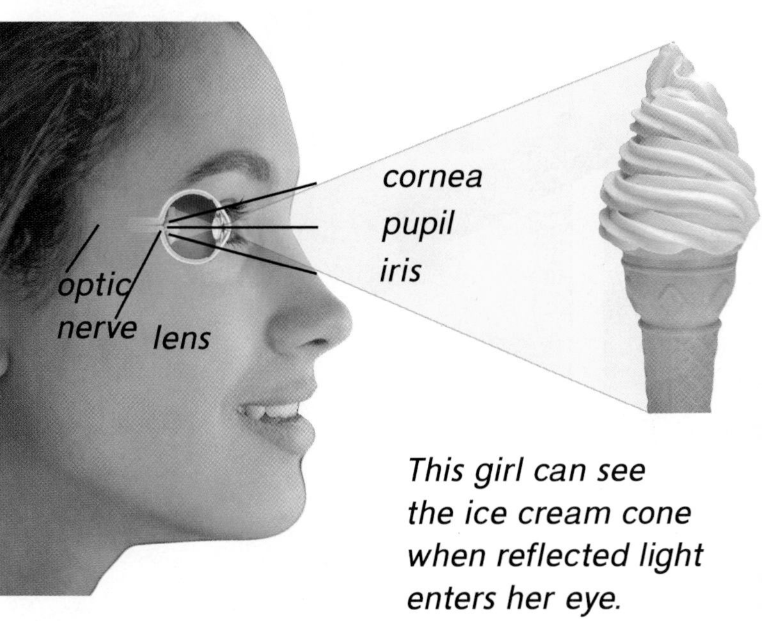

This girl can see the ice cream cone when reflected light enters her eye.

Did You Know?

You actually see things upside down. The refracted and reflected light inside your eyeballs produces upside down versions of the image you are seeing. Your brain combines the information from both eyes and flips it right side up again so you sense things the way they really are.

What Could I Be?

Optician

People's eyes do not always focus light well enough for them to see clearly. An optician designs and fits glasses and contact lenses to help correct poor vision. Learn more about the work of an optician in the Careers section.

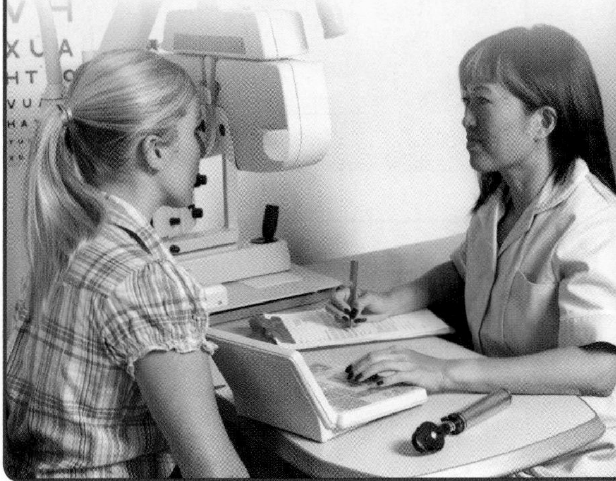

Future Career

Robotics Engineer Do you wonder what the future will be like? Robotics engineers are paving the way with their technology. These engineers use computers to research and design ways to use robots. Then they build and test their designs. Robotics engineers have to understand how systems work, especially electricity and machines. Their work with robots has been used in many different things. From surgery, to space vehicles, to toys, robots are becoming a reality!

(bkgd) Janka Dharmasena/iStock/Getty Images; (inset) Fstal/age fotostock

Engineering

Introduction to Technology and Engineering

Technology and Engineering

Animals use the world around them to make their lives easier. A bird will build a nest from sticks and leaves. A fox will burrow into the ground to make a home. Humans also change the environment to help us survive. We change the world around us to make our lives easier and more comfortable.

Technology is the use of science to solve problems. A pencil and paper is a technology that has been around a long time. A smart phone and tablet computer are newer technologies. Some technologies use simple tools. Others use many science ideas and tools at the same time.

People are always looking for new ways to solve problems using technology. When people get sick, technology may be able to help. Technology can help us build strong and sturdy homes. Every part of our lives can be changed or helped by using technology.

Did You Know?

Technology does not refer just to electronics and gadgets. You use many technologies around you that you do not even notice. For example, carpet does not just decorate a room. It is a technology that helps quiet a room and keep it warm. Modern carpet is made of material that cleans easily.

Nearly everything in your home is a type of technology.

Engineering is the process of planning and designing technology. Engineering and technology are closely related ideas. The goal of engineering is to come up with better technology. *Engineers* are people who use science knowledge to invent technology that solves problems. They combine ideas and materials to invent solutions.

The idea of engineering is not new. Even people who lived long ago had to solve problems. They had to find food and stay warm. They came up with ideas to help them do this. They used rocks and sticks to make tools. They used animal skins to keep warm. If they made mistakes, they changed the process or tools they used the next time.

Engineers do the same thing today. People live in modern cities and have useful inventions because of engineers. People travel around the world and into space because engineers have invented and improved technologies.

Complicated technology comes from adding up many smaller ideas over time.

Even a simple tool is an example of technology.

Finding Technology and Engineering

Everywhere you look, you can find technology and engineering at work.

Technology is used to plant, harvest, and produce our food.

Food

The earliest farmers used simple tools to hunt and plant food. Today, farming is done on a large scale. Machines are used to plant and harvest crops. Factories with machines package and store the foods. At home, we use technology to prepare and cook foods. Ovens, microwaves, and blenders are just a few technologies we use in cooking. Forks and spoons are technologies too.

Technology makes modern homes sturdy and safe.

Shelter

Humans have come a long way since the shelters of early cave people. Sturdy building materials stand up to fierce storms. Plumbing and electricity make our homes comfortable. Technology can even help us clean our homes.

Transportation

A cart with wheels is one of the earliest transportation inventions. Over time, boats, trains, cars, and planes changed transportation. Engineers built roads and bridges. They designed railroad crossings. They found ways to make thousands of planes take off and land safely around the world each day.

Transportation technology moves people by land, sea, and air.

Communication

First came the invention of paper and then the printing press. Later came the telegraph, telephone, and television. Now a phone with wires seems old fashioned. Engineers figured out how to use waves in the air to send and receive messages. Today we can carry complex gadgets to keep in touch.

Communication technology has changed greatly over the last 100 years.

Medicine

Today's medical technologies save many lives. Medicines are made of chemicals that help people. Technologies can be used to see inside the body. Medical devices can help people live with disabilities. Eyeglasses are an example of medical technology. They help solve a problem for people who have trouble seeing.

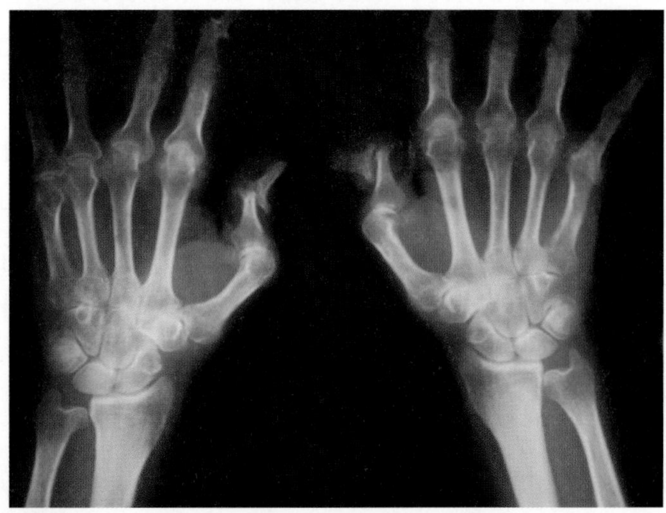

Medical technology helps people live longer.

Recreation

Technology can solve many important problems, but it can help us have fun too! To stay happy and healthy, people need a break from work. Movies use technology to entertain us. Sports equipment and toys are made with technology. Art supplies, bikes, and video games are more examples of technology.

The things that entertain us are built using technology.

Science and Math Connections

Science and math are both big parts of engineering. Engineers use science to understand the properties of materials they work with. They must think about how the materials act under different conditions.

Think about an engineer making a new kind of bicycle. There is a lot of science at work. The metal used must be sturdy. It must be strong enough to hold the weight of the rider. It must also be lightweight so the bike is not too heavy to pedal. The parts that hold the bike together must also be strong and lightweight. A bike must be able to be outside in all kinds of weather. The bike's wheels must hold a rubber tire in place. The air in the tire cushions the rider against bumps on the ground. All of these things are made possible through science.

A metal called aluminum alloy is light, stiff, and strong. It is often used to make bicycles.

Math is also very important in engineering. In making the bicycle, engineers use exact measurements. Then those exact measurements must be used by factories building the bikes. If the math is not exact, the engineer may design a bike that could be unsafe. It will not work the way it should. The factory must use the exact measurements, too. If they do not, they may not be able to put the bike together.

Engineers also use math to communicate ideas. They use it to collect and display data about how their designs work. A line graph or bar graph can show how different materials react in different conditions.

Math tools are also used in engineering. Some tools are as simple as a ruler. Others are more complex. Tools may measure size, mass, heat, light, or sound. Engineering uses very precise math tools to make sure information is accurate.

Engineers use computers to create models using math data they have collected.

Engineering Design Process

Engineers follow the same steps each time they set out to solve a technology problem. The steps are called the *engineering design process*. The process is sometimes called a loop. That is because steps are repeated until the solution works well.

Identify a Problem First, engineers think about what problem needs to be solved. Do they want to make a safer kitchen tool? Do they want to help lower school energy bills? They may want to make buildings that can stand up to intense storms. Sometimes an engineer might just come up with a great idea.

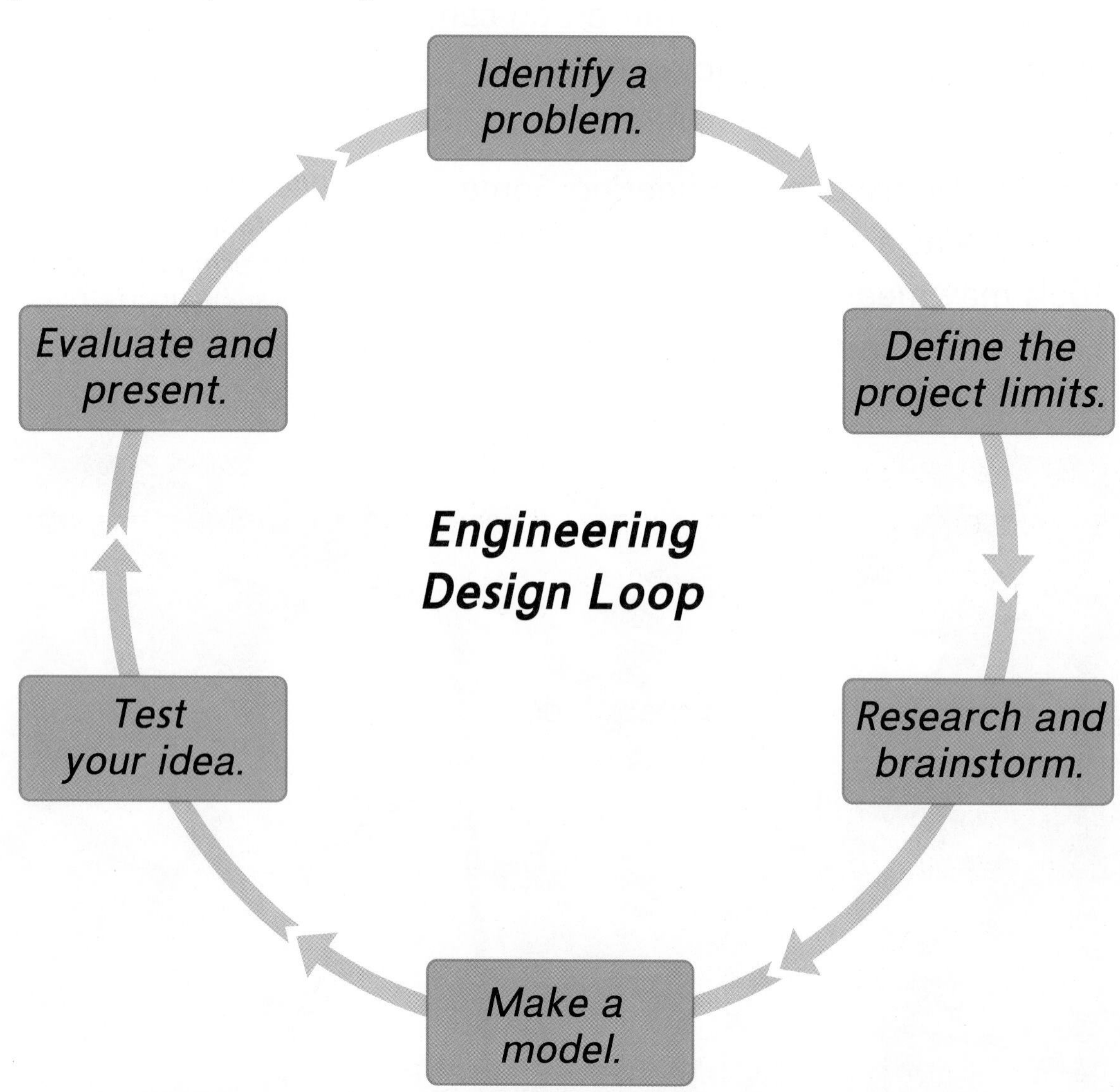

If you wanted to design the fastest skateboard, you would try several shapes before deciding which one worked best.

Define the Project Limits Next, engineers think about how big their project will be. How much time will they have? How much money will they have? Who will help them? What materials can they use? Where will they work? These questions are important for an engineer to answer before starting a project.

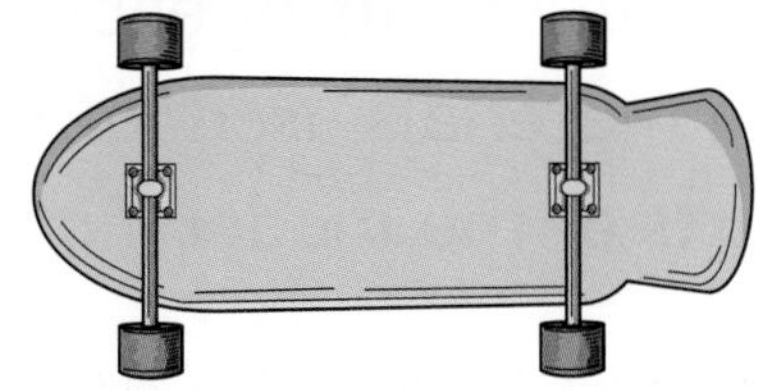

Research and Brainstorm There are many ways to solve a problem. Engineers research how others tried to solve the problem in the past. Then they brainstorm new ideas for solving the problem.

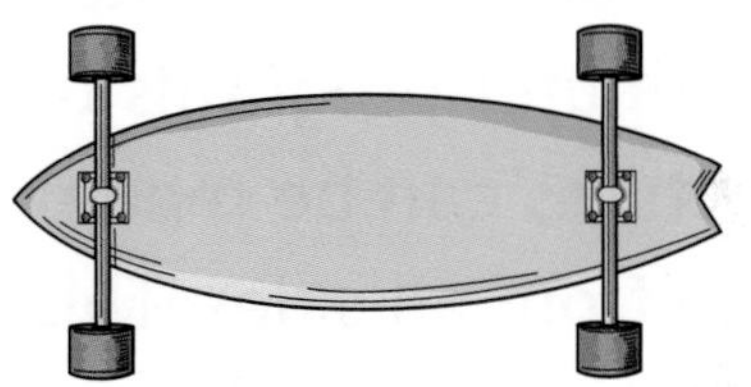

Make a Model Next, an engineer puts together a model of his or her tool or technology. This model may become a big loop in the process. The engineer may have to try many different models to find the one that works best.

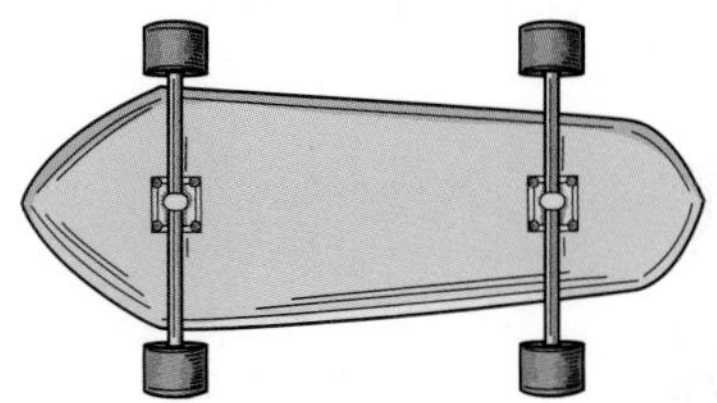

Test Your Idea Once a model is made, it must be tested. Testing may show that more work must be done. Engineers ask: Does the design work? Does it solve the problem? Can it be made better? If the idea does not work, a new model or idea is needed.

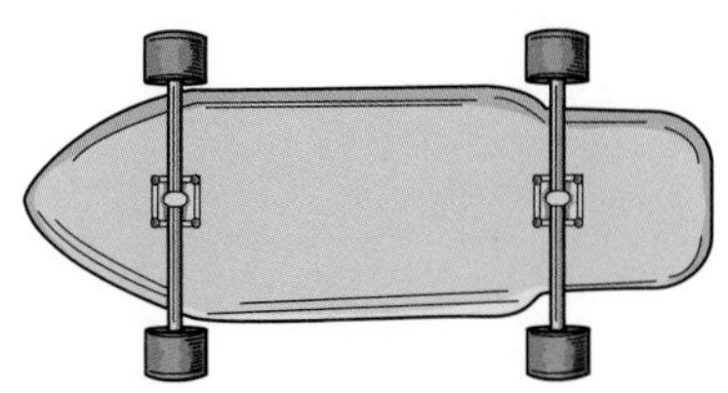

Evaluate and Present Once a working model is made, the design can be presented. Other people might want to test the new design. People can evaluate how well the new design works.

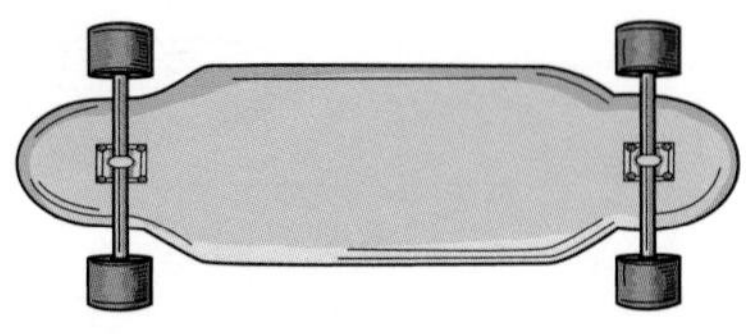

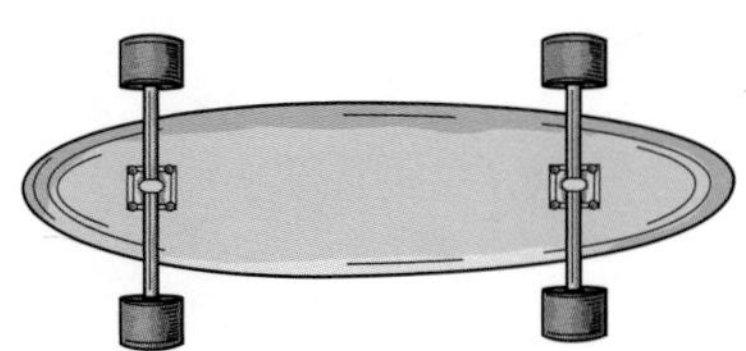

Design and Modeling

Making sketches and models is a great way to communicate ideas. Sketches show what a model should look like. A sketch should have as many details as possible about a plan. A sketch might include measurements of each part of the design. It might label the materials used for each part. Arrows can show how parts of the model move.

Sketches may be drawn on paper, but could also be made on computers. Scientists and engineers use computer programs to make a very exact sketch. These pictures can be made to show what an object will look like from all sides. This ability helps engineers see many details at once. The computer can keep track of changes to measurements.

Computer drawings let designers zoom in and out and view the model from different sides.

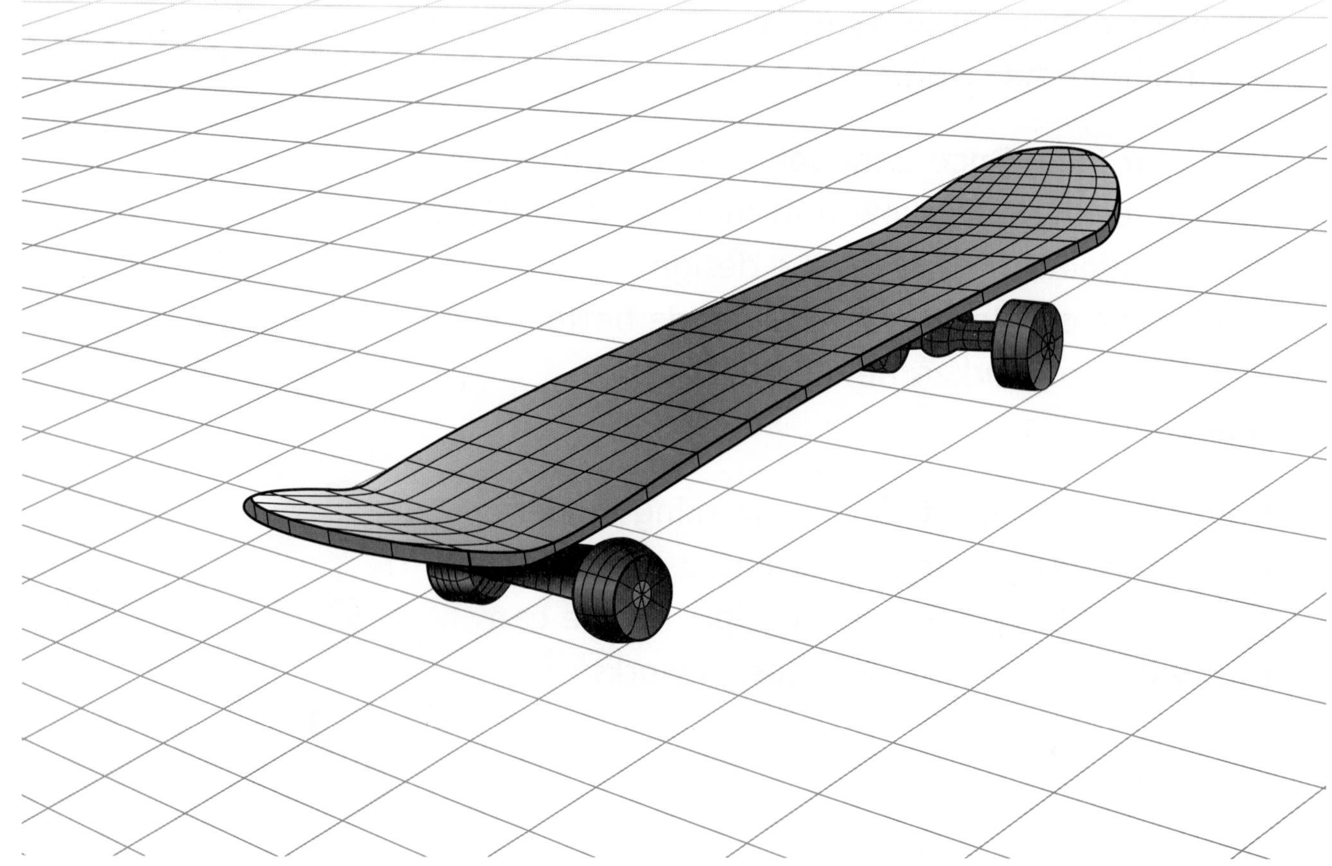

After a detailed drawing, an engineering designer usually makes a model. Imagine building an actual bridge before building a model first. How would you know what size to make the materials and how to connect them?

A *scale model* shows an accurate relationship between all the parts of the model. Math and computers help with calculations of the sizes and shapes of parts.

Suppose a scale model of a bridge is 100 times smaller than the real bridge. Everything else about the bridge is the same. This model helps engineers know how the bridge will look and how it will work.

A scale model might be used to see how much weight the bridge can hold too. The model may be designed to hold 100 times less weight than the real bridge. Models help engineers test their designs.

This model lets engineers test how the full sized bridge will handle forces.

Systems and System Thinking

A *system* is a collection of parts that affect one another. The way the parts work affects how the whole system works. If one wheel on your bike is low on air, it makes the whole bike harder to pedal. Engineers think about how the parts of their technology work together. They also think about how their solutions fit into bigger systems.

Natural Systems The natural world around us is full of systems. The solar system is made of the Sun, planets, moons, and other objects. The human body is also made of systems. A system of muscles and bones work together to help us move.

Artificial Systems Many systems are made by people. A school is a system of grade levels and classrooms. Your route to school is a system of roads and sidewalks. Engineers design systems, too. The Internet is a complex system of connected computers.

This picture shows a scale model of the human skeletal system. You can study it to see how the parts work together.

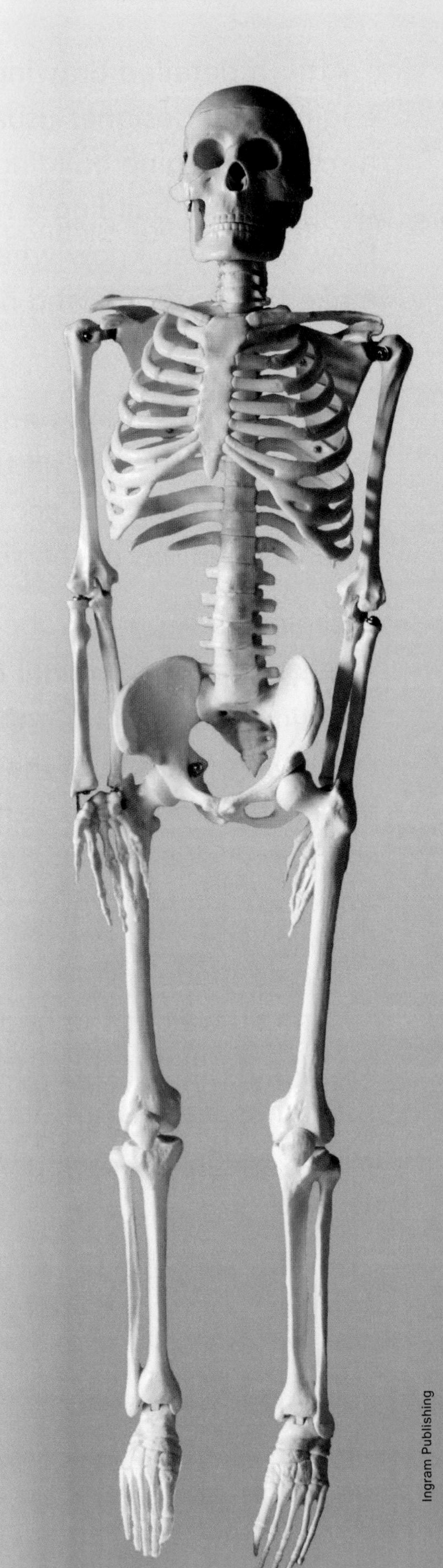

Thinking About Systems

Engineers know that the environment affects how their technology will work. They know the technology also affects the environment. For technology to work the way it is supposed to, engineers must think about how several systems affect one another. The purpose of technology is to solve problems. Good technology must solve one problem without creating other problems.

Engineers also think about the relationship between *inputs* and *outputs*. Input is the power, energy, or information put into a machine or system. Output is the power, energy, or information delivered or produced by a machine or system. They ask questions such as, What do we have to *put in* this solution to make it work? Will the outcome be worth the effort and material we put in? If the answer to the second question is yes, then the technology will probably be put to use. Engineers do not just guess at the answer, though. They gather *feedback*. Feedback is information about how a solution is working. Ideas from feedback provide some of the input for the next version of a design.

Did You Know?

The solar system can affect cell phones! Explosions at the surface of the Sun release bursts of radiation. The radiation can cause many types of electronic devices to fail.

NASA/GSFC/SDO

Materials and Fabrication

Before making products, engineers must find out a lot about the materials they want to use. Metals, fibers, plastics, and ceramics are types of materials. Engineers test the strength of materials. They consider if a material bends. They understand how it reacts to heat and light. They think about whether it conducts electricity well.

Engineers work in labs to test different materials. They put them under strain and stress. Machines pull and bend them. Engineers will not use materials that do not stand up well in their tests. For example, new materials for firefighter tools must be fireproof and lightweight.

Some engineers invent new materials. Chemicals may be added or taken away from materials we already use. They may make the materials at different temperatures. This process changes the properties of the material.

The rubber that makes up a bike tire is flexible but does not stretch enough to fall off the wheel. The wheel rim is rigid to hold its round shape.

Glow Images

Fabrication is the making of a material or product for many people to use. When engineers make a product, the final step is to *manufacture* it. Manufacturing means making something from raw materials by hand or with a machine.

Sometimes a mold is made for liquids to be poured into. When poured into the mold, the liquid will harden into the shape of the mold. Other products are formed from materials at high temperatures. The material may be stretched or pushed into shape by machines. Then the parts cool and harden.

Some products have pieces that are joined together. A car is made of many separate parts. Some materials need to be treated with special chemicals. Car seats may go through a liquid that keeps them from easily catching fire. Finally, some materials must have a special finish. Paint on the outside of a car protects it from the weather.

Factories are places where fabrication and manufacturing take place.

Types of Engineering

Civil Engineering

There are many kinds of engineers. A *civil engineer* works on projects that are used by the public. These projects could include stadiums, airports, bridges, and tunnels. It includes roads, canals, and waste disposal systems. Some of the most amazing structures around the world were designed, planned, and built by civil engineers.

Civil engineers must consider many things when building large public projects. They must choose long-lasting materials that do not harm the environment. They must choose a stable place to build. The structure must allow existing wildlife to survive. Public safety is the most important thing for civil engineers to think about.

Civil engineers work on projects and structures that will be used by the public.

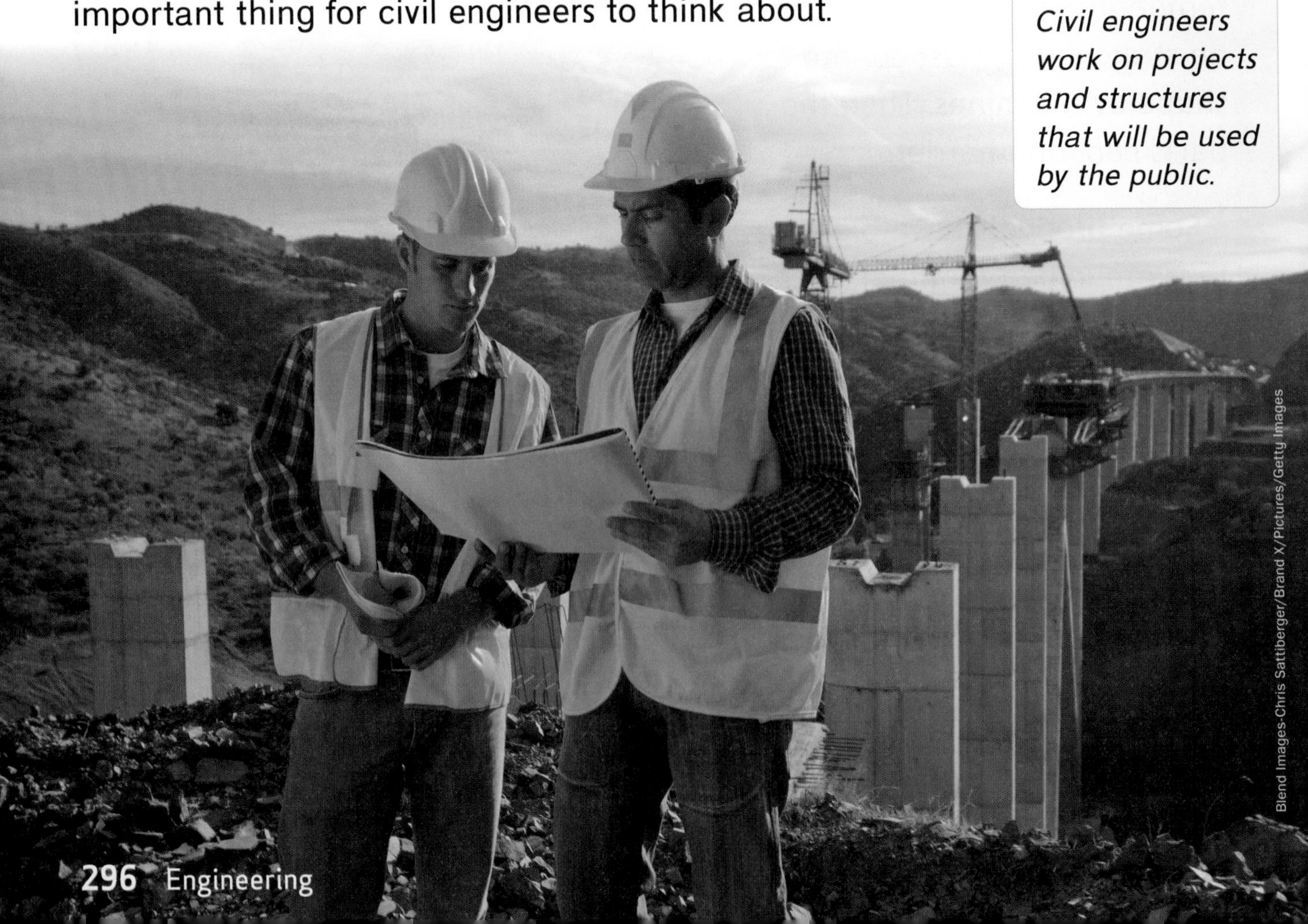

Blend Images-Chris Sattiberger/Brand X/Pictures/Getty Images

Electrical and Electronics Engineering

An *electrical engineer* works on projects that use electric current. Electrical engineers plan the systems that bring power to your neighborhood and home. They design the systems that power amusement park rides, factories, and schools. They must make sure the right amount of electricity is generated and sent through wires. Safety is a big concern for electrical engineers. They may build backup systems, called generators. That way if power goes out in an area, people can still have the electricity they need.

An *electronics engineer* designs projects that use electronic signals. These projects include smart phones, video games, recording devices, and even cable television.

Electrical and electronics engineers must understand how electricity moves. They build in safety features for the devices they design. For example, many appliances include residual current devices, or RCDs. RCDs monitor the current flowing in an appliance and shut off the appliance if the current does not flow properly.

Electronic devices use electricity, whether it comes from batteries or a wall outlet.

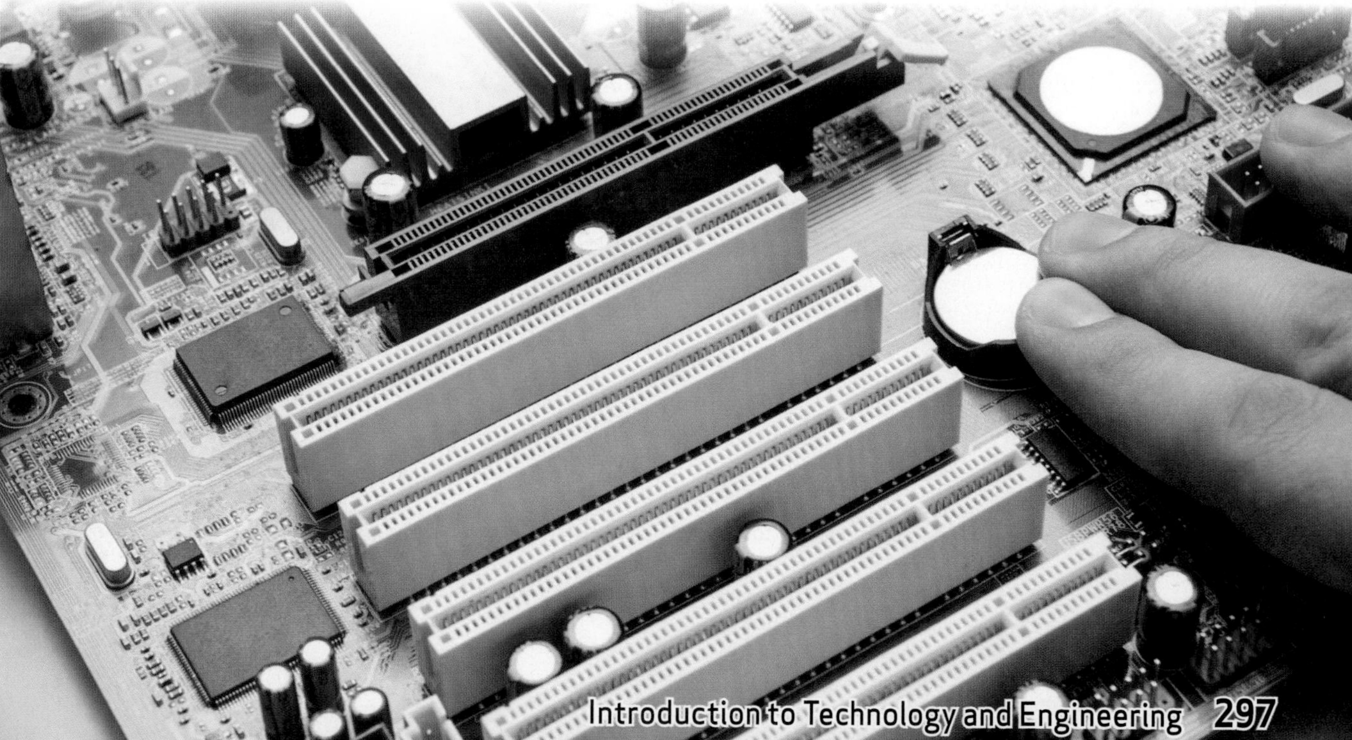

Piotr Adamwicz/iStock/Getty Images

Types of Engineering - Mechanical Engineering

Mechanical engineers design projects that have moving parts. Mechanical engineers may work with other kinds of engineers to design a complete project. For example, mechanical engineers plan the moving parts of a wind turbine. They find out the number of spins per minute needed for the turbine to make energy. They know how much spinning might make the system overheat or stop working. The movement of the propellers is important to the way the whole system works.

Moving parts are used in many industrial machines. Mining and drilling for oil require such machines. Mechanical engineers understand the force needed to do different jobs. Simple machines play a big part in many mechanical design plans. Mechanical engineers understand how mechanical energy is changed to other kinds of energy.

Mechanical engineers design products with moving parts.

Lars Ruecker/Getty Images

Manufacturing Engineering

Manufacturing engineers are involved in the process of making products. Today, every part of every airplane and computer goes through a manufacturing process. Almost every piece of clothing and every train, car, and bicycle part is made in a factory. Manufacturing engineers constantly improve the process of making things with less waste.

Manufacturing engineers might look at a model made by another engineer or designer. Then they find the best process for making thousands or millions of that item. They help determine which materials work best. They consider the cost and how easy the material is to work with.

A manufacturing engineer might come up with a new kind of machine that can make items faster. They might suggest changes to the product that make it easier or faster to produce. Most items you buy in the store are the result of manufacturing engineering.

Manufacturing engineers come up with better ways to produce goods quickly and with less waste.

Glow Images

Types of Engineering - Biomedical Engineering

Biomedical engineers solve problems related to medicine and human health. They make products such as artificial limbs. They develop systems that make a doctor's job easier or safer. They develop new products that help save lives. Biomedical engineers might even make products that make the body more comfortable when sitting or typing for a long time.

Biomedical engineers design machines that take pictures of the inside of the body. Doctors can see diseases inside people's bodies. They can see babies before they are born.

Biomedical engineers must understand a lot about how the human body works. They must know what artificial materials can be used inside the body. This fact is important in designing artificial body parts, such as hip replacement joints. Biomedical engineers also design new instruments used in surgery to reach specific parts of the body.

Word Study

Bios is the Greek word meaning "life," and *medicus* is the Latin word for "physician," or doctor. Biomedical engineers develop ways to help doctors improve people's lives.

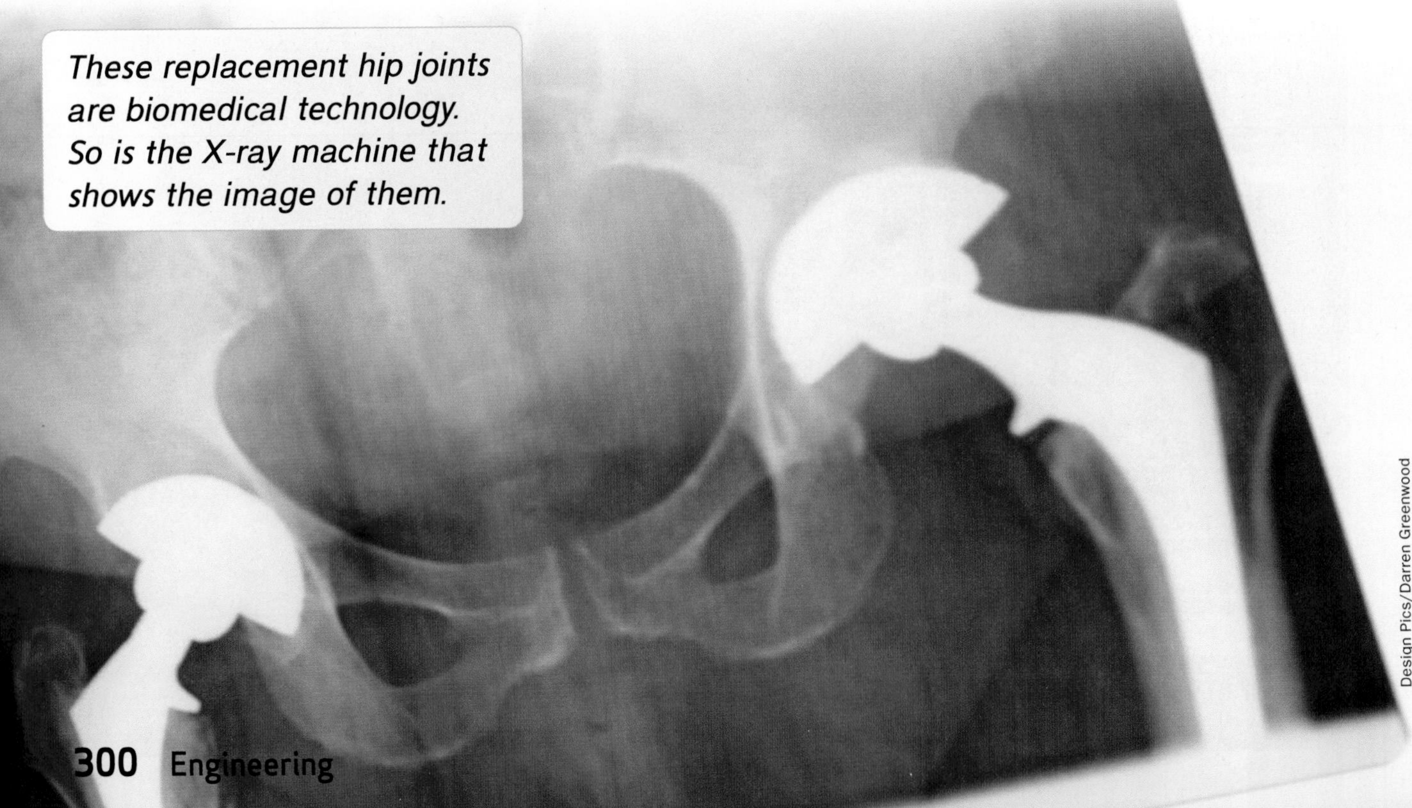

These replacement hip joints are biomedical technology. So is the X-ray machine that shows the image of them.

Design Pics/Darren Greenwood

Chemical Engineering

Chemical engineers design products or processes that use chemicals. These products and processes can include dyes in clothes and additives in foods. They include plastics used to package toys. Chemical engineers spend a lot of time testing chemicals to make sure products are safe.

Chemical engineers also design processes to make materials. For example, chemical engineers probably worked on the process of making the paper that you write on. They develop the chemicals used in transforming trees into pulp. There are also chemicals in the inks used in the paper. Even the packaging uses chemical processes.

Chemical engineers also help make products so they can be recycled. Even the recycling process is designed and improved by chemical engineers.

Chemical engineers help people save money and help the environment by developing materials that can be recycled into new products.

Engineering Teams and Skills

When engineers work together, amazing technologies take shape. Different kinds of engineers use their expert knowledge on the same project. Large public projects need many people to work on them.

Engineering Teams

When an engineer has a good idea that can help solve a problem, he or she will make a plan. Different people might help the engineer in different parts of the design process. Not all of the people who work with engineers have gone to school for engineering. Many have other skills that engineers need. Large building projects need many construction workers. They need electricians. They need people who lay pipes and mix concrete.

Another person on an engineer's team might be a drafter. That is someone who is very good at making technical drawings. Many people who produce these drafts use computer programs to do their work.

Working on an engineering project takes teamwork.

Pixtal/age fotostock

Some people who work on engineering projects are researchers. The very beginning stages of the design process involve research. Scientists will find out how other people have tried to solve the problem. They might research what technologies have worked on similar problems. This information is useful during a team's planning stages.

The development stage of an engineering project involves research too. Some scientists will test how materials react to different conditions. They might research the cost of materials and where they can be found.

Other people on engineering teams might work with the public. They ask people to test how early models of a product work. The people give useful feedback to the engineers. The engineers can find out what needs to be improved. Then they can work on making those changes.

Engineering teams may work with the public to test products.

Patrick Ryan/Digital Vision/Getty Images

Engineering Teams and Skills - Engineering Skills

An engineer uses many skills in his or her work. Engineers need science skills. A biomedical engineer must know a lot about the human body. A student studying biomedical engineering will take many biology classes. To make a new kind of medical tool takes knowledge about how different chemicals work together in materials. If the tool has moving parts, knowledge of mechanics is needed. If the tool uses electricity, then the engineer must have electrical knowledge. Even when specialized teams work together, the leading engineer needs to understand the whole system.

Engineers must also understand the risks involved in what they are doing. Engineers often lead teams of workers. For example, an electrical or chemical engineer must not ask a team member to do something dangerous.

Engineering teams use many science skills.

Teamwork is a big part of working on an engineering project. Team members must communicate well with one another. In the manufacturing process alone, there are many people working on different tasks. Any change in the process probably affects everyone. It must be communicated to the right people.

From the start of a project, engineers present ideas to team members. They must describe their plans to people who will help them research. They must explain how they see the project turning out. Speaking and writing are important skills for an engineer. Even when a project is finished, they must explain how it will help people. They may need to explain how new technologies work.

Making decisions is also a part of the engineering process. When and how should a design be improved? When is more research needed? These are important decisions that engineers must make as part of their job.

Engineers must communicate their ideas to others.

kristian sekulic/E+/Getty Images

Engineering Challenges of the Future

In the future, many more engineers will be needed. Our world is using more and more new technology every day. Every new technology opens doors to new inventions and designs.

Energy

Many engineers today are trying to find new ways to produce the energy we use. Our world is using more and more energy every day. The old ways of producing energy are hurting the environment. Fossil fuels burned to make electricity and heat will run out. Engineers try to find ways to use renewable resources, rather than coal or oil, for energy. That means finding way to use energy from wind, water, or the Sun. New technologies can cost a lot. Engineers try to find ways to make these technologies less expensive and to make them easy for people to use.

Engineers face the challenge of making renewable energy available to everyone.

What Could I Be?

Marine Energy Engineer

The constant motion of waves and tides is a renewable energy source that can be harnessed. Marine energy engineers study the oceans and design floating and underwater generators that produce electric current. Learn more about marine energy engineering in the Careers section.

Huge turbines like this can be placed underwater to capture the power of ocean tides.

(l) Pedro Castellano/Getty Images; (r) Michael Roper/Alamy

Infrastructure

The way a city or system is organized is called its *infrastructure*. The roads, bridges, and tunnels that help people move from town to town are parts of an infrastructure. The electrical and telephone lines in a city or town are also infrastructure. So are water lines and sewers. Engineers try to make infrastructures work as well as possible. They try to make them more modern.

Some engineers work on improving old infrastructure. They replace parts of old roads or bridges. They may replace old underground cables with new technologies. Improving old infrastructures can help us keep up with changing technology.

Other engineers design brand new infrastructures. Computers are faster and more powerful than ever. Finding new ways to send information quickly means that a modern cable infrastructure is needed. With each new technology, new needs develop. New problems arise. Engineers work to solve these problems.

A computer system can be part of a school's infrastructure.

Engineering Challenges of the Future - Global Health Concerns

Biomedical engineering is one of the fastest growing fields in science. People are living longer now than ever before. These longer lives come partly from the work of the medical field. However, there is still a lot to improve around the globe. In some areas of the world, there is a shortage of clean water. People can become sick from unclean water. Chemical and biomedical engineers can help. They work on technologies to bring nations around the world what they need. These needs include clean water and ways to get rid of waste properly.

Engineering is important for controlling and curing diseases. Health technologies are needed to keep people safe from disease. These systems can mean adding nutrients to some foods we eat. These added nutrients can help keep people healthy. Manufacturing new medicines can also help protect people around the world from diseases and improve their quality of life.

Medicines, like other products, are manufactured in factories.

Nickondr/iStock/Getty Images

Globalization

As new technologies are invented, more people have access to places far away. Today it is easy to get on a plane and fly all the way across the world. The Internet makes it simple to communicate with people thousands of miles away.

The process of interacting on a worldwide scale is called *globalization*. New technologies make globalization possible. New technologies can instantly translate words from one language to another. People can buy goods online and have them shipped right to their homes.

Globalization can create new challenges, however. Private information on the Internet must be kept safe. Engineers who work with computers help solve this problem. Countries have rules about how people can cross their borders. Technology helps manage that process, as well.

Technology makes it possible for us to buy something from the other side of the planet and have it delivered to our door, without ever talking to another person.

Siri Stafford/Photodisc/Getty Images

Information Technology

Information technology is the use of systems that send, store, or receive information. The telephone was one of the first electronic information technologies. It was invented in the late 1800s. It took time to set up new telephone systems. First, miles of wires needed to connect from city to city. Soon, telephone cables went into people's homes. Over time, technologies improved. Sounds were clearer over the wires. Telephone lines today connect cities and towns all around the world. Most homes have telephone wires in nearly every room.

Cellular Communication

Technologies continue to change, however. Today many people use cell phone technology that does not even use wires. Instead of sending messages through wires, cell phones send signals through the air. Signals are sent from tower to tower. Newer cell phones send signals that are faster and stronger. Cell phone technologies are improving every day, the way wire phone technologies did before.

Information technology changes quickly.

George Doyle/Getty Images

Computer Engineering

Computers are also changing. These changes make information technology a growing field of science. Some people who work on computer technologies are called *software engineers.* They develop new computer programs. These new programs help people work in faster and easier ways. Software engineers may help companies come up with programs that meet their special needs. A doctor has different computer needs than a banker has.

Software engineers also develop technologies that people can use at home. The applications, or apps, you may use on a cell phone or computer tablet are made by software engineers. They use the design process to design and develop apps. Then they test them and make sure they work with different types of computers. Like other technologies, computer technologies change often. A software engineer working today may use different systems than someone who did the same job a few years ago.

What Could I Be?

Computer Hardware Engineer

Computer companies seem to release smaller, lighter, and faster versions of their products every few months. Computer hardware engineers design and develop new hardware, but they also regularly look for ways to refine their designs. Learn more about computer hardware engineering in the Careers section.

Computer hardware engineers design the computers you use. Software engineers design the programs on the computers.

Information Technology - Data Storage

Not long ago, storing digital information was a big challenge for engineers. Computers the size of an entire room were used to hold data. The information those computers held would seem like a very small amount today. A smart phone today can hold entire libraries of books and music.

The Internet is a global network linking millions of computers. Data for the Internet are stored and controlled by computer systems called servers. These servers can hold large amounts of digital data.

Like all technologies, Internet technologies are changing. Data from many servers around the world can be stored in an Internet system called the cloud. That makes it possible for even more information to be stored and then found and used again.

In the future, data storage methods will change again. Engineers play a large part in designing and planning new computer systems.

All the data in all the books you see here can be stored in a tiny device such as this USB drive.

Space Technology

Technology and science have come so far that they have even reached out into space. Space technology is another growing field of science.

A satellite is an object in space that moves in an orbit around a moon or planet. Engineers have sent over a thousand satellites into Earth's orbit. These satellites send and receive signals from the ground. Satellites help us communicate around the globe. We send out television signals using satellites. These signals allow one event to be viewed by people all around the planet at once, just as the event is happening.

Engineers do more in space than just communicate with Earth for entertainment. Engineers design robots to explore other planets. Their work is helping us examine new places, such as Mars. In the future, the work of engineers will continue to help us learn and solve problems.

People can even live in space on space stations created by engineers.

Future Career

Emergency Medical Technician (EMT)

Do you like your days to be full of action? Do you enjoy helping others? An emergency medical technician has a fast-paced and very important role in society. They make sure that people who have been in an accident or have a health emergency get to the hospital as swiftly and as safely as possible. EMTs need to know advanced first aid, CPR, and know how to think quickly and critically about how to best help the person in need. These experts study throughout their career to make sure they are up-to-date on how to care for others when they are in need of help.

Human Body

Nutrition and Nutrients

Page 316

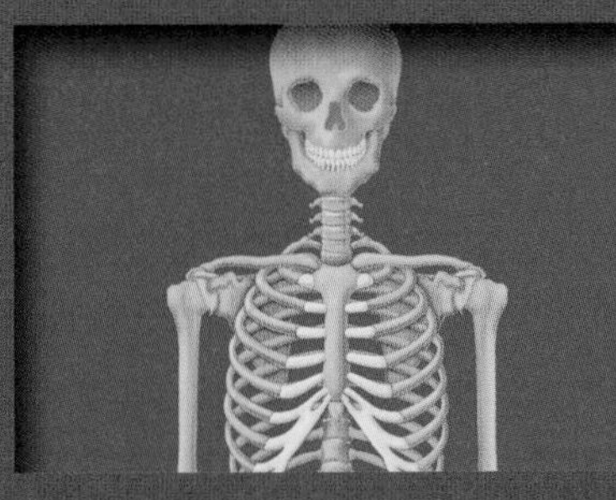

Body Systems

Page 318

Nutrition and Nutrients

The human body needs energy to live and grow. Nutrients are the materials in foods that help the body grow and get energy. Healthy foods supply the nutrients the body needs. The six kinds of nutrients are described on these pages.

Carbohydrates

Carbohydrates are the body's main source of energy. Some carbohydrates, such as potatoes, breads, and pastas, provide the body with long-lasting energy. Sugars are also a type of carbohydrate. Sugars are found in fruits and sweets. They provide the body with short-term energy.

Vitamins

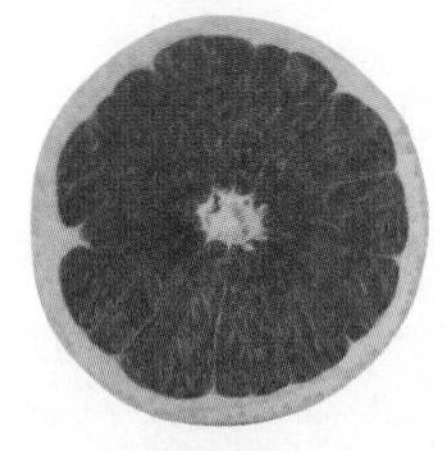

Vitamins perform specific jobs in your body. Different vitamins have different jobs. For example, Vitamin B_6 helps your body get energy from food. Eating a variety of nutritious foods helps ensure that the body gets the vitamins it needs.

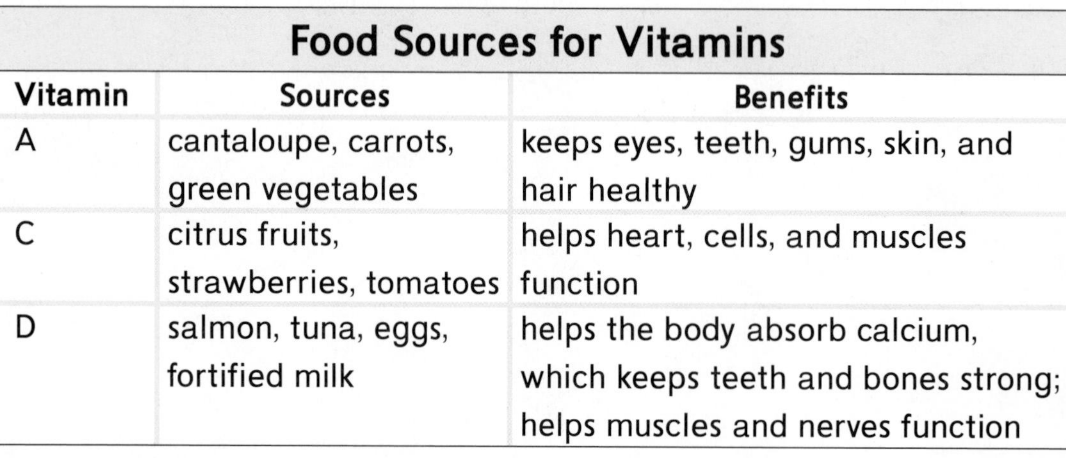

Food Sources for Vitamins		
Vitamin	**Sources**	**Benefits**
A	cantaloupe, carrots, green vegetables	keeps eyes, teeth, gums, skin, and hair healthy
C	citrus fruits, strawberries, tomatoes	helps heart, cells, and muscles function
D	salmon, tuna, eggs, fortified milk	helps the body absorb calcium, which keeps teeth and bones strong; helps muscles and nerves function

Minerals

Minerals help the body grow and function properly. Like vitamins, different minerals have different jobs. Potassium, found in bananas and sweet potatoes, helps nerves and muscles work. Magnesium, found in nuts, seeds, and whole grains, helps control muscles and nerves.

Food Sources for Minerals		
Mineral	**Sources**	**Benefits**
calcium	yogurt, sardines, fortified cereal, kale	builds strong teeth and bones; helps with muscle function
iron	meat, beans, fish, fortified cereal, spinach	builds new blood cells and helps them function
zinc	meat, fish, eggs	helps the body grow; helps heal wounds

Fats

Fats help the body store vitamins and energy. Your body uses the energy in fat when it does not get energy from other nutrients. The body needs only a small amount of fats. Fats are found in meats, eggs, butter, cheeses, oil, and nuts.

Proteins

Proteins help build and repair cells, and they help bones and muscles grow. Proteins help the body fight diseases and heal wounds. Foods high in proteins include milk, eggs, meats, fish, nuts, tofu, and soybeans.

Water

Water is necessary for life. About two thirds of the body is made of water. Water helps the body remove wastes and carries nutrients to the cells of the body. Water also helps prevent the body from getting too hot.

(t-b) ©Foodcollection; Comstock Images/Alamy; Natalya Chumak/Hemera/Getty Images; ©Photodisc/Getty Images; lumusphotography/iStock/Getty Images; Pixtal/age fotostock

Body Systems

The Skeletal System

A body system is a group of organs that perform a specific job. The bones of your body make up your skeletal system. The *skeletal system* supports the body and gives it shape. Together with the muscles of the body, the skeletal system allows the body to move. Bones also store minerals such as calcium. Blood cells are made inside the bones in tissue called bone marrow.

There are 206 bones in the human body. Each bone has a specific job. For example, the rib cage protects the body's organs such as the heart and lungs. The skull bones protect the brain. The long bones of the legs support the body's weight.

A joint is a place where bones meet. There are three main types of joints.

- Immovable joints form where bones fit together too tightly to move. The bones of your skull meet at immovable joints.
- Bones that meet at partly movable joints can move a little. Ribs are connected to the breastbone with these joints.
- Bones that meet at movable joints can move easily. Knees, hips, shoulders, and elbows contain movable joints.

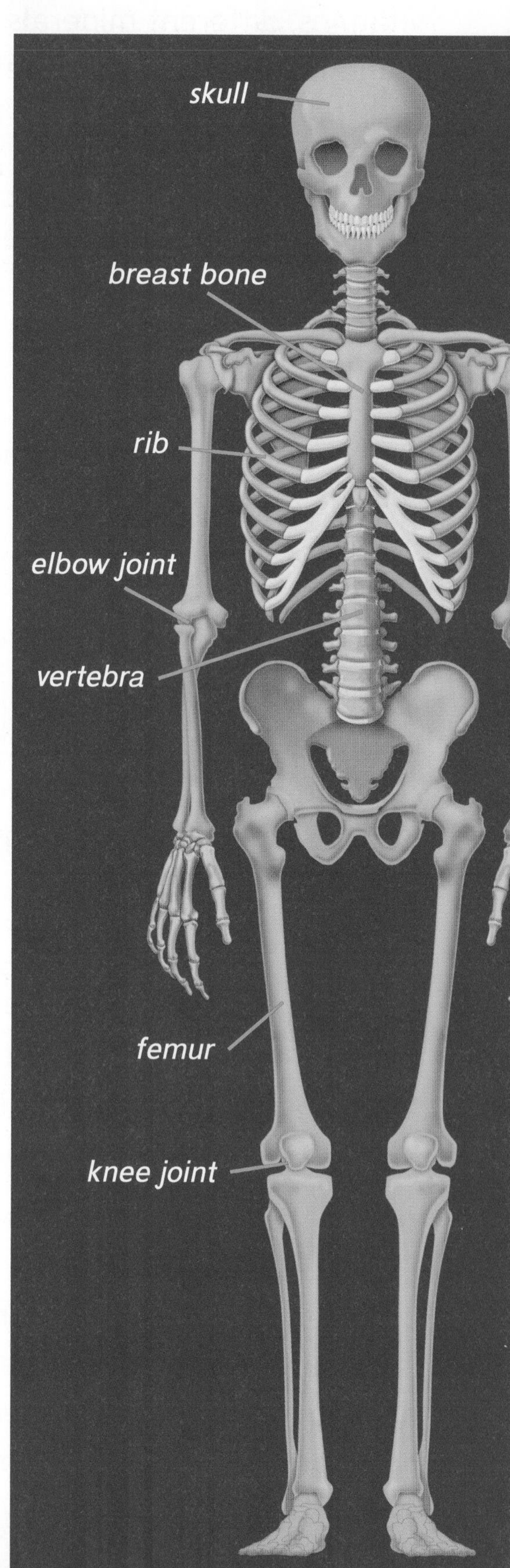

The Muscular System

The *muscular system* is made up of all the muscles in your body. A muscle is a body part that causes movement.

Most muscles are skeletal muscles. Skeletal muscles are attached to the bones and skin. Tough, cordlike tissues called tendons attach these muscles to the bones. Tendons and muscles work together with the skeletal system to move the body. Skeletal muscles usually work in pairs. When you want to move, the brain sends a message to a pair of muscles. One muscle contracts, or gets shorter. It pulls on the bone. The other muscle relaxes to let the bone move. You can control your skeletal muscles.

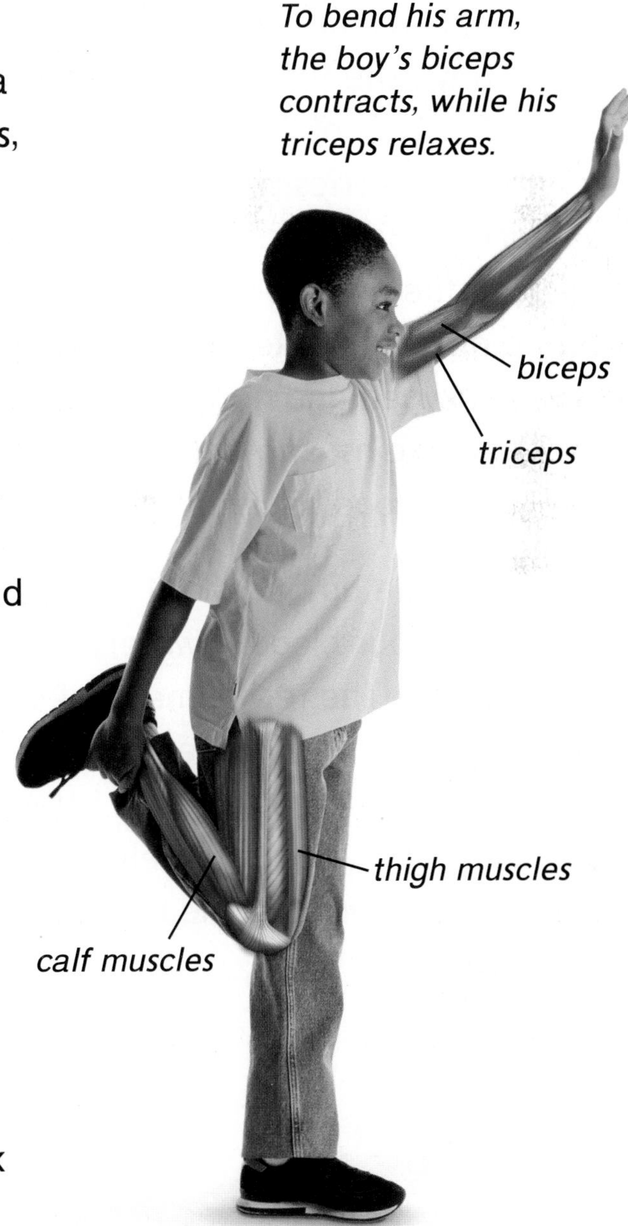

To bend his arm, the boy's biceps contracts, while his triceps relaxes.

The heart is made up of another kind of muscle called cardiac muscle. You cannot control these muscles. They keep your heart beating day and night.

Smooth muscles are found in the walls of many of the body's organs. They line the blood vessels and help blood move through the body. One of these muscles in the chest helps you breathe. Smooth muscles in the digestive system contract and relax, moving food through the body. Like cardiac muscle, smooth muscles work without you thinking about them.

The Circulatory System

The *circulatory system* is the body's transport system. It sends oxygen and nutrients to the cells of the body. Blood vessels, blood, and the heart make up the body's circulatory system.

The heart is an organ about the size of a fist. It beats about 70 to 90 times each minute. Its muscles pump blood through tiny little tubes, called blood vessels. One type of blood vessel is an artery, which carries blood away from the heart. In the lungs, blood picks up oxygen. In the digestive system, blood picks up nutrients. Arteries carry blood rich with oxygen and nutrients to the body's cells. Veins, another type of blood vessel, carry blood back to the heart.

Blood is made up of several parts. Over half of your blood is watery liquid called plasma. The rest is made of cells. Red blood cells carry oxygen to all parts of your body. White blood cells attack and kill invading germs. Platelets help clot the blood and stop the bleeding from a wound.

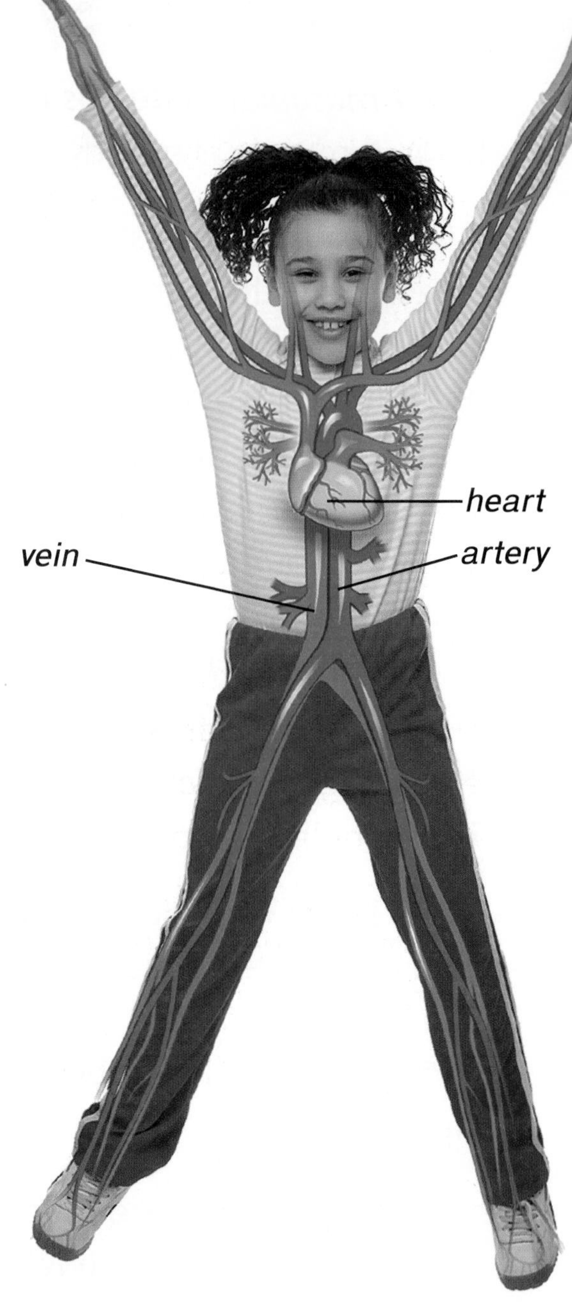

red blood cells

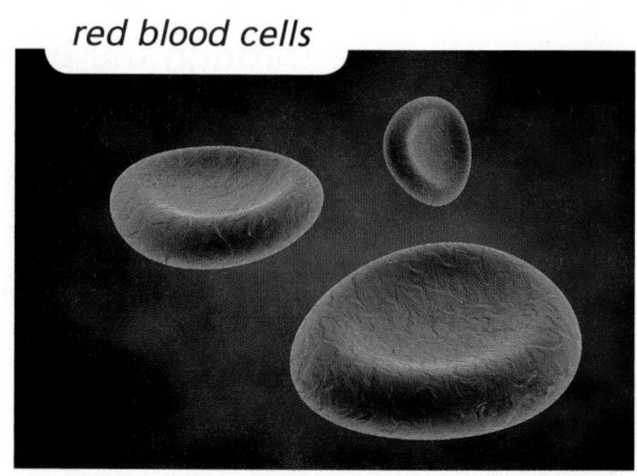

The Respiratory System

The *respiratory system* moves gases in and out of the body. All the body's cells need oxygen in order to work properly. You take in oxygen from the air when you breathe in, or inhale. As you inhale, a muscle called the diaphragm contracts. This muscle allows the lungs to expand. Air flows in through the mouth or nose and into the trachea. From the trachea, air flows into two bronchial tubes. Each tube leads to a lung. Inside the lungs, the bronchial tubes branch off into smaller tubes called bronchioles.

As air flows into the bronchioles, it reaches millions of tiny air sacs. It is in the air sacs that oxygen is picked up by red blood cells. The red blood cells also release carbon dioxide into the air sacs. As you exhale, the diaphragm relaxes. The lungs deflate and push carbon dioxide and other waste gases out of your body through your nose and mouth.

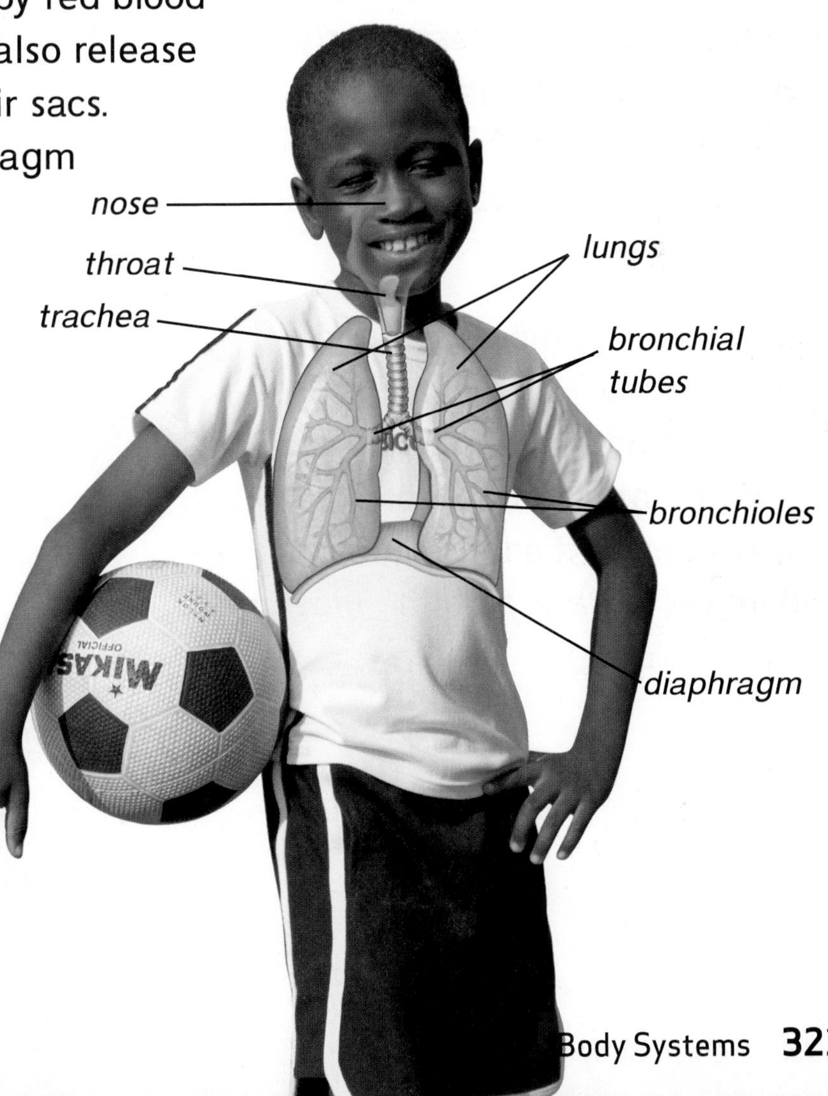

The Digestive System

The *digestive system* breaks food down into nutrients that can be used by the body. Digestion begins in the mouth, where food is torn and mashed into small pieces. Saliva moistens the food and helps it travel smoothly when you swallow. The food travels down your esophagus and into your stomach. Once food is in the stomach, muscles contract and mash it. The stomach is filled with acid that helps break down the food. Soon the food turns into a thick liquid.

The food then moves into the small intestine, a coiled, narrow tube that is about 6 meters (20 feet) long. This intestine is where most of the nutrients in the food are absorbed into the blood. Blood carries these nutrients to all parts of the body.

Next, the food moves into the large intestine. Here, water is absorbed from the remaining food. The food that is left is waste. It passes through the large intestine and leaves the body.

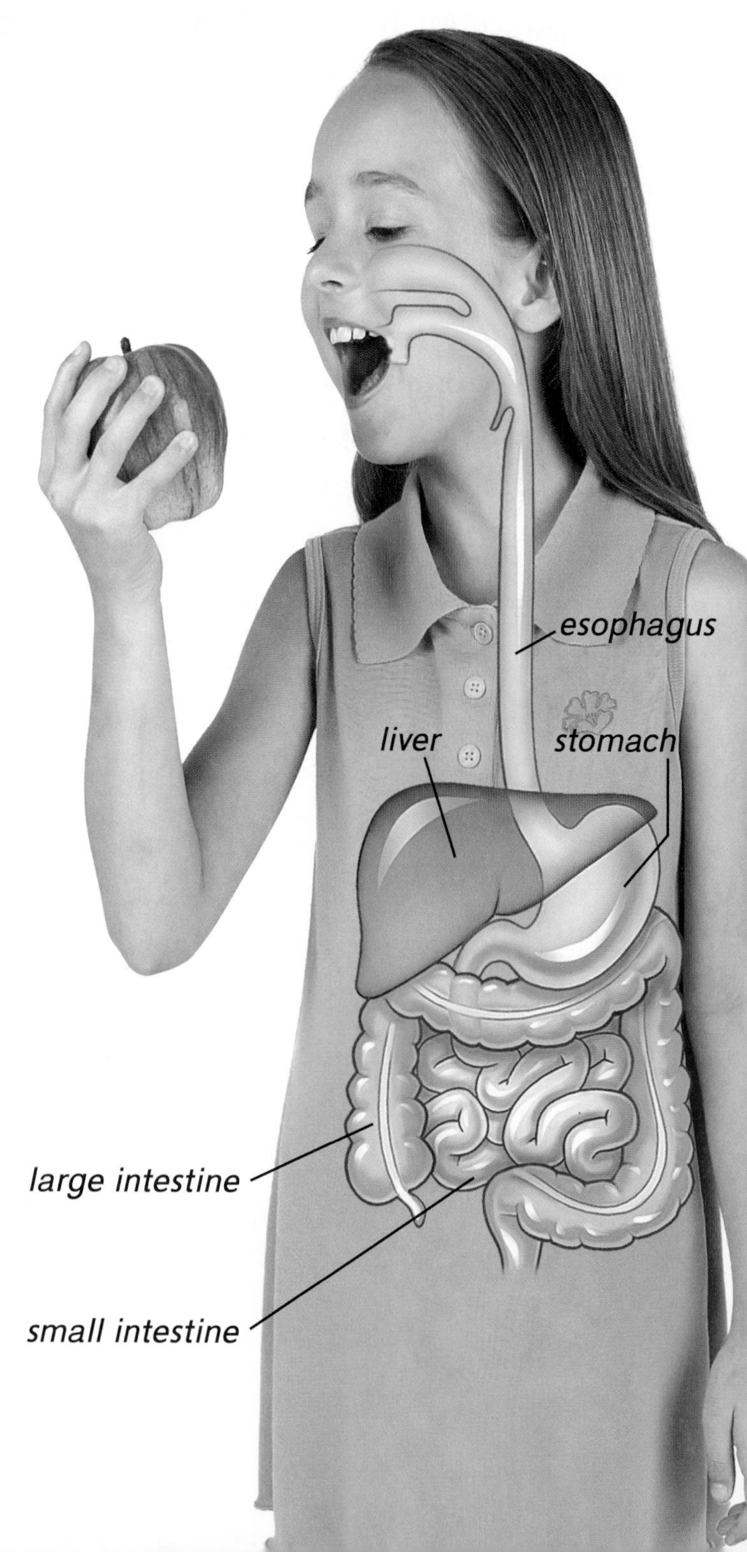

The Excretory System

The *excretory system* removes wastes from the body. Wastes are materials that the body does not need, such as extra water and salts. The liver, kidneys, bladder, and skin are organs of the excretory system.

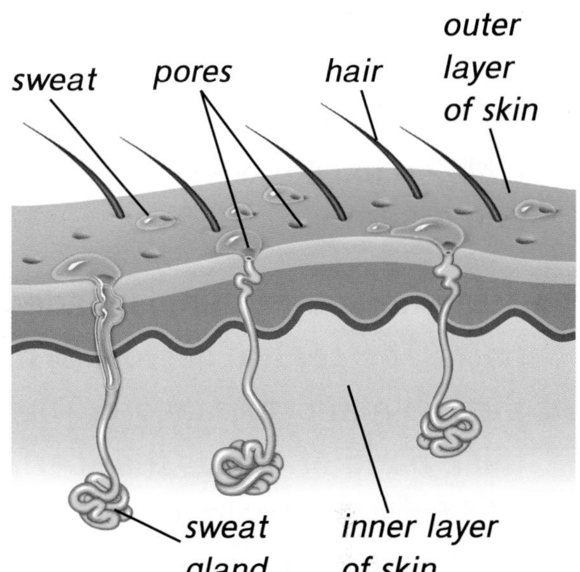

Liver, Kidneys, and Bladder

The liver filters the wastes from the blood, changing these wastes into a chemical called urea. The urea travels from the liver to the kidneys, where the urea is changed into urine. Urine flows through two long tubes to the bladder. It is stored in the bladder until it leaves the body through the urethra.

Skin

Skin helps rid the body of wastes when a person sweats. Sweat glands are found below the surface of the skin. Sweat is made up of water and minerals the body does not need. When a person sweats, wastes are released from the sweat glands through the outer layer of skin. Sweating cools the body and helps it maintain an internal temperature of about 37°Celsius (98°Fahrenheit).

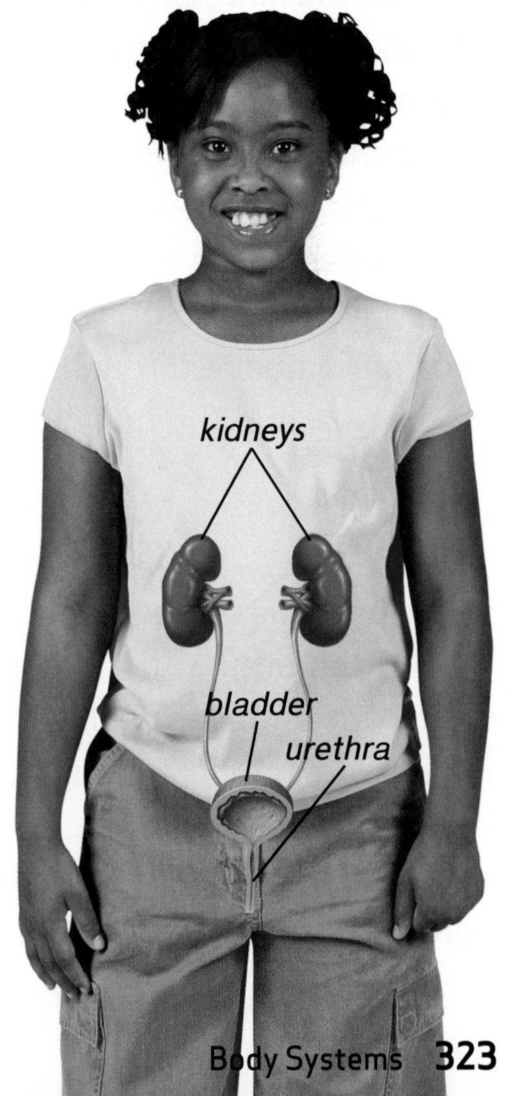

The Nervous System

The *nervous system* is the control center of the body. It takes in and responds to information. It also controls many jobs of the other body systems, such as balance and movement. It allows a person to think, feel, and dream.

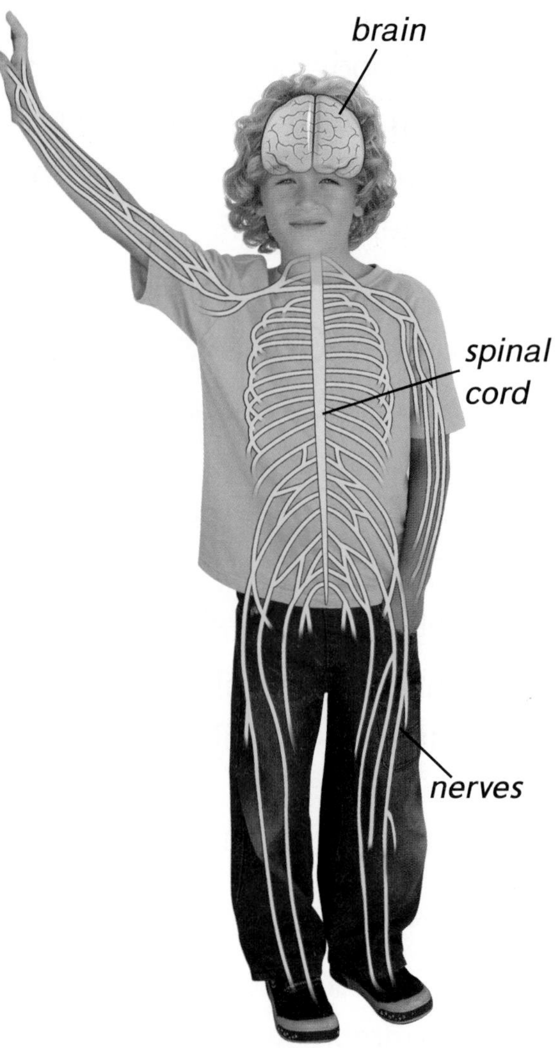

The nervous system is made up of two main parts. The central nervous system includes the brain and spinal cord. The spinal cord is a thick band of nerves that carries messages to and from the brain. All the other nerves make up the peripheral nervous system. Nerves branch from the spinal cord to all parts of the body. Nerves are made up of cells called neurons. Neurons receive information from cells of the body. They pass this information to the brain through the spinal cord. When the brain receives the information, it decides how the body should respond. It passes this new information back through the spinal cord to the nerves, and the body responds.

The Brain

The brain has three main parts. The largest part of the brain is the cerebrum. It stores memories and helps control information it receives. The cerebellum helps the body balance and directs skeletal muscles. The brain stem connects the brain to the spinal cord. It controls heartbeat and breathing.

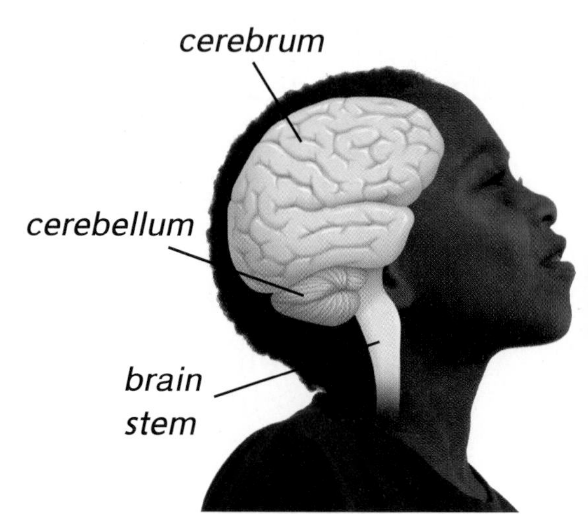

The Five Senses

Different nerves in the body take in information from the environment. These nerves are responsible for the body's senses of sight, hearing, smell, taste, and touch.

Sight Light passes through the pupil, a small opening in the eye. Cells in the eye change the light into electrical signals. The signals travel along the optic nerve to the brain.

Hearing Sound waves enter the ear and reach the eardrum. Cells in the ear change the sound waves into electric signals. The signals travel along the auditory nerve to the brain.

Smell Chemicals in the air enter the nose. Certain cells in the nose send the information along a nerve to the brain.

Taste The tongue is covered with between 2,000 and 4,000 tiny bumps called taste buds. Each taste bud can sense sweet, sour, salty, savory, or bitter tastes. Taste buds are also found at the back of the throat and in the nose.

Touch Nerve cells in the skin give the body its sense of touch. Signals are sent to the brain through the spinal cord. The brain can tell hot, cold, soft, hard, and other touch sensations.

(t-b) KYU OH/Getty Images; Purestock/SuperStock; MariaPavlova/Vetta/Getty Images; Ingram Publishing; Comstock Images/Alamy

The Immune System

The *immune system* helps the body fight off germs. Germs can cause disease and infection. The body has defenses that help prevent germs from entering. The skin, saliva, and tears work to kill germs and keep them from entering the body.

Sometimes germs make their way into the body. White blood cells help find and kill germs quickly before they cause illness. White blood cells, which are part of the blood, travel through blood vessels and in lymph vessels. Lymph vessels, which are similar to blood vessels, carry fluid called lymph. Many white blood cells are made and live in lymph nodes. The cells filter out harmful materials that flow through the lymph nodes.

Sometimes the white blood cells do not kill all the germs. When germs reproduce in your body, they can cause illness. Even then, the immune system continues to work to kill and remove the germs.

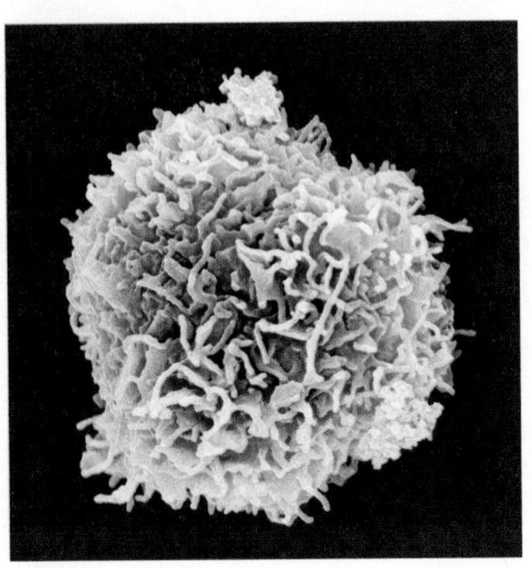

This picture shows how a white blood cell looks through a microscope.

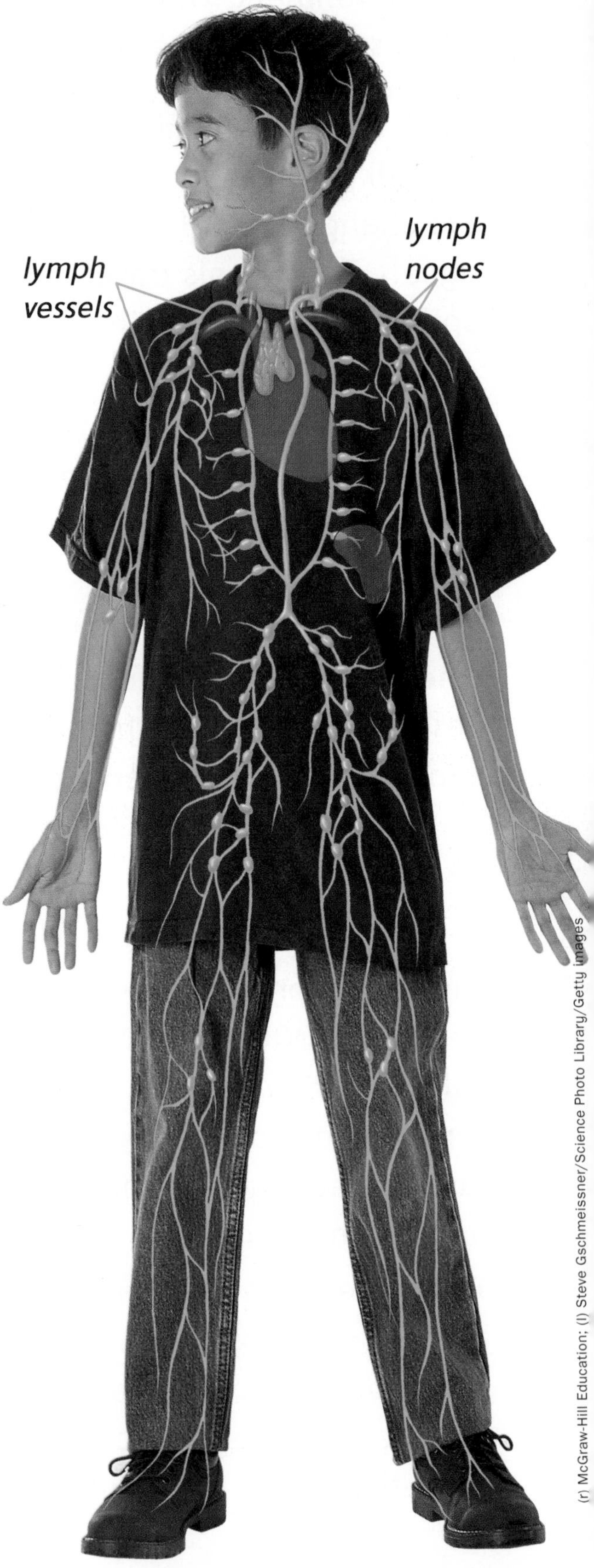

lymph vessels

lymph nodes

Viruses and Bacteria

One type of germ that makes people ill is a *virus*. Viruses can cause colds or the flu. There is no cure for these illnesses. Instead, the body must fight off the infection on its own. However, vaccinations protect against many illnesses caused by viruses such as the flu, measles, and chicken pox.

Viruses are not considered living things. They need to be inside living cells to reproduce. As they reproduce, they take away nutrients and energy from the cell. They also produce harmful materials that make the body itch or cause high fevers.

A *bacterium* is another type of germ that can cause illness. Bacteria are one-celled organisms that live outside the body on most surfaces. Some bacteria can have a harmful effect on the body. Other bacteria are helpful. For example, some bacteria help the body digest food.

Help your body defend itself against germs.

- Eat a variety of healthful foods.
- Be active. A fit body is better able to fight germs.
- Get a yearly checkup.
- Get plenty of sleep. Students your age need about 10 hours of sleep each night. Get more rest when you are ill.
- Do not share cups or utensils.
- Wash your hands, especially before eating and drinking.

Cold viruses as seen through an electron microscope

E. coli *bacteria as seen through a microscope*

Scrubbing your hands with soap removes many germs that can cause illness. Wash your hands for as long as it takes to hum "Happy Birthday" twice.

If soap and water are not available, use hand sanitizer. Put the product in the palm of one hand. Rub it all over both hands and fingers until your hands are dry.

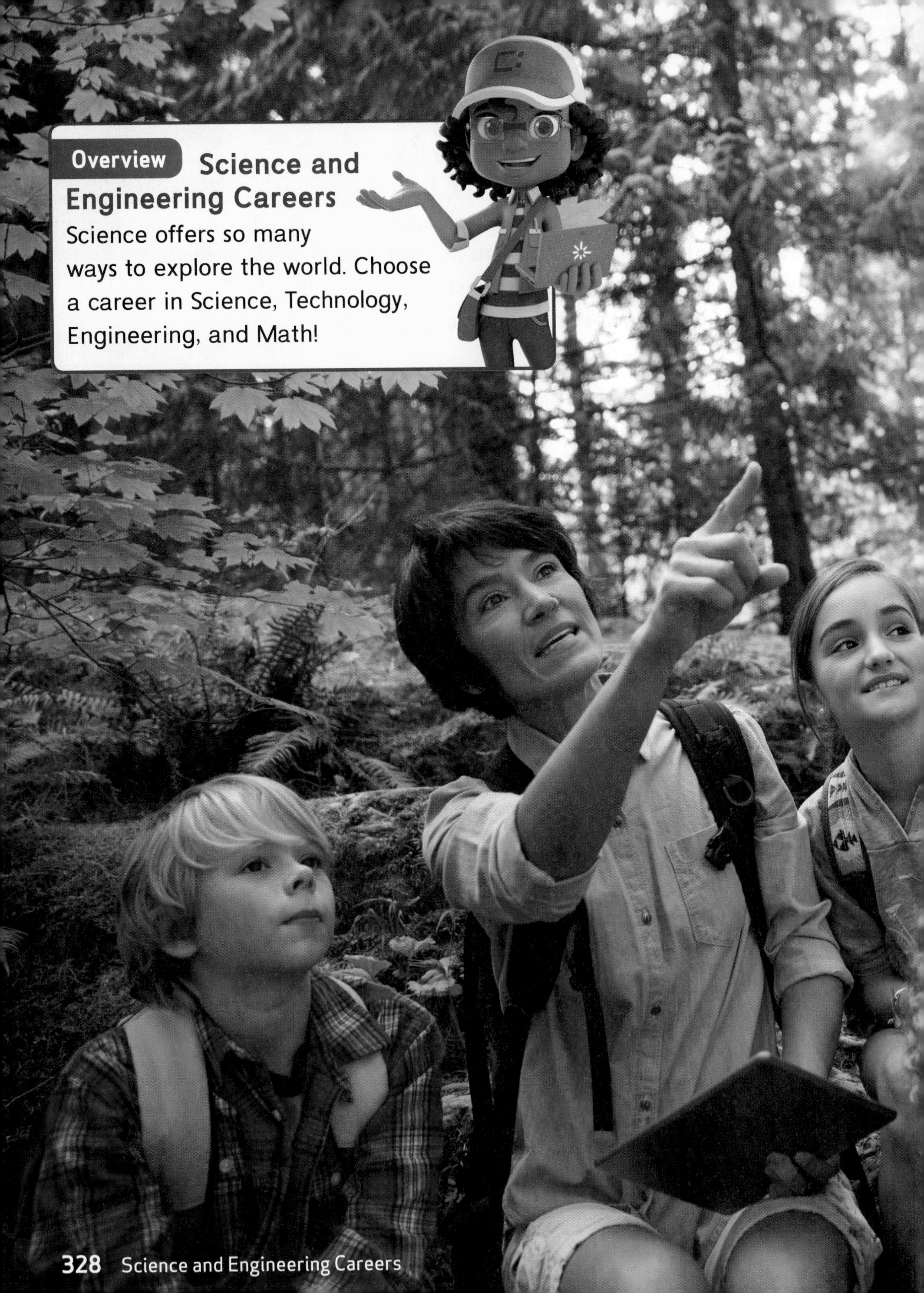

Overview **Science and Engineering Careers**

Science offers so many ways to explore the world. Choose a career in Science, Technology, Engineering, and Math!

©Hero Images Inc./Corbis

Science and Engineering Careers

Life Science Careers

Some bacteriologists work to keep our water supplies safe. Others study helpful and harmful bacteria in plants and animals.

Bacteriologist

Are you interested in seeing germs up close? Do you want to learn how they affect living things? If so, a career in bacteriology may be for you. Bacteriologists study tiny organisms called bacteria in order to help keep people safe from harmful diseases. Bacteriologists may work for a university, a hospital, a drug company, or for the government. These scientists usually have a college degree in microbiology. Many bacteriologists eventually earn a doctorate degree.

Plant Biologist

If you are fascinated by the world of plants, then a career in plant biology could be an option. Also called botanists, plant biologists study plants ranging from the very small watermeal to the giant sequoia tree. Plant biologists may grow plants in a lab. They may work in the field to find out how plants are affected by their environment. Biology, botany, chemistry, and math are important college courses for future botanists.

Some plant biologists study variations in a single species of plant.

Horticulturist

Horticulturists study the best ways to grow and care for plants. Some of these scientists work with farmers to grow fruit and vegetable crops that resist disease. Others may work in an arboretum, which is a garden devoted to trees. If you want to become a horticulturist, you should take college courses in landscape design and plant biology, and horticulture classes such as plant breeding and crop production.

Understanding how plants adapt to different conditions is one of the jobs of a horticulturist.

Geneticist

Geneticists study genes, or chemicals that determine traits, in living things. They try to determine how parent organisms pass on these traits when they reproduce. A geneticist studying plants may work to produce the sweetest, juiciest apple. Another may work to develop a rose that is the perfect shade of pink. A geneticist who studies people may try to determine how genes can cause certain diseases to be passed on. Most geneticists have a degree in biology or medicine along with a degree in genetics. Math and statistics are other courses that many geneticists find useful.

Geneticists study models of the chemicals that carry genetic information.

Marine Biologist

Do you enjoy hunting for sand crabs or observing fish below the ocean's surface? Perhaps you should consider a career as a marine biologist! The word *marine* means "ocean." Marine biologists not only study saltwater organisms but also work to protect them. They can work in many places, from a laboratory to a marine sanctuary, and even in the ocean itself. Biology, physics, chemistry, oceanography, and zoology are just a few of the subjects studied by students who wish to become marine biologists.

Marine biologists collect data and study ocean life up close.

Ecoinformatics Specialist

Every day, thousands of scientists around the world are observing and gathering information on everything from plants and animals to weather patterns. An ecoinformatics specialist uses technology to organize, analyze, and manage these data. They also help scientists find ways to share their data with others in their field of study. People with careers in this very new field must have a strong background in science and engineering as well as computer information systems. If you enjoy science, math, and working with computers, then this may be the career for you!

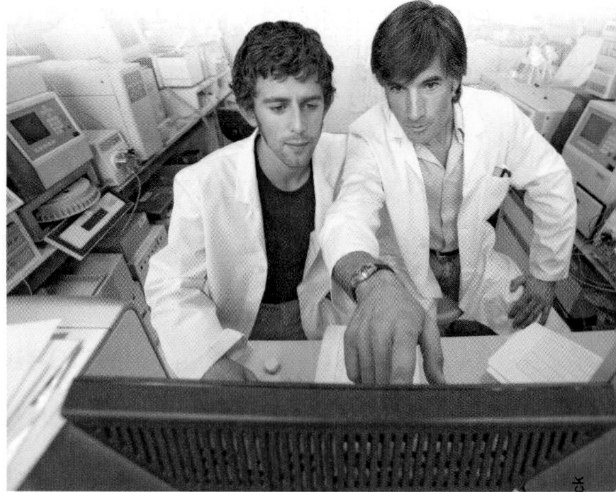

Crunching data and finding ways to communicate the information are the main goals of an ecoinformatics specialist.

Desalination Engineer

Our world is made up mostly of water. Only about three percent is fresh water. The rest is salt water. Desalination is the process of removing salt and other minerals from water, making it safe for humans to drink.
A desalination engineer designs equipment and processes that companies can use to filter salt out of salt water without using large amounts of energy. If you are interested in becoming a desalination engineer, you will need a strong background in the Earth and life sciences as well as engineering, chemistry, and physics.

Desalination engineers provide fresh water needed by people and industry.

Energy Auditor

Do you care about conserving energy to help our Earth? More and more home and business owners want to reduce the amount of energy they use. Energy auditors help them do just that. Energy auditors use technology to detect leaks, inspect insulation, check air ducts, and examine heating and cooling equipment. Auditors use their findings to make suggestions that can help people save energy and money. Most auditors have a background in engineering as well as special training.

An energy auditor helps identify air leaks that can cost homeowners money.

(t) Thinkstock/Stockbyte/Getty Images; (b) Jupiterimages/liquidlibrary/Getty Images

Earth Science Careers

Trail Designer

When you enjoy a hike in the woods, do you ever think about how the trail formed? Many people assume that trails form over time through the action of animals, water, and natural events. Some trails are formed naturally, but many trails are designed and built by people. A trail designer designs trails that can be used for hiking, off-road bicycling, horseback riding, and other kinds of recreation. They design trails for everyone from landowners to the National Parks Service.

The slope of the land, the proposed use of a trail, and the ecology of an area all factor into trail design.

When planning a trail, a designer has to consider more than just how the trail will be used. Erosion patterns, rock and soil types, weather, and native plants and animals all affect a trail's design. Trail designers must help people enjoy nature while also protecting the land. People interested in a career in trail design might study art and design, Earth sciences, drafting, forestry, and engineering.

Many trails are built and maintained by the National Park Service for people to enjoy.

Volcanologist

Do you like to solve mysteries? Many volcanologists view their work as similar to the work of detectives. Volcanologists are scientists who study how and why volcanoes erupt. They monitor active volcanoes, examine rock samples from dead volcanoes, and research how volcanic eruptions affect living things. Volcanologists must have a strong background in geology, which is the study of rocks. They must also have a good understanding of Earth science, chemistry, and math. It takes most students eight to nine years to become a volcanologist.

The temperature of lava ranges from 650 degrees Celsius to over 950°C (1200 degrees Fahrenheit to 1750°F). A heat-resistant suit protects this volcanologist taking the temperature of a lava flow.

Seismologist

Do you wonder what makes earthquakes start and stop, or what controls their timing? If so, then seismology might be the field for you. When rocks beneath Earth's surface move, they release energy in the form of seismic waves. Seismologists study these waves as they move through and around Earth. They monitor seismic activity that is happening now and research activity that happened in the past. Seismologists use this information to help engineers design buildings that can better withstand the shaking caused by large earthquakes. In college, future seismologists study geology, geophysics, math, and computer science.

Seismographs record the waves produced by earthquakes. Seismologists study these waves to look for patterns.

Meteorologist

Are you are interested in the weather? Do you enjoy using computers and other kinds of technology? If so, then meteorology might be the field for you. Meteorologists observe weather, record and analyze data, and make predictions about the weather. They use many different tools in their work, including anemometers, thermometers, and barometers. They also use satellites, watercraft, aircraft, and weather balloons. Students who wish to become meteorologists study atmospheric sciences, calculus, physics, and computer science.

Meteorologists use data gathered by various tools to make predictions about the weather.

Climatologist

If you are more interested in weather patterns that happen over a long period of time, then you might want to consider a career as a climatologist. Climatologists research temperature, wind, and precipitation patterns over many years. They study evidence of how climates have changed throughout Earth's history. They also investigate variations in climate and how the climate affects living things, including the crops we grow for food. They help engineers determine which materials are best to use in certain climates. Climatologists need a strong background in math, physics, Earth sciences, and computer technology.

Climatologists study evidence of past climates recorded in ice.

Paleontologist

If you are interested in dinosaurs and other forms of prehistoric life, a job in the field of paleontology may be for you. Paleontologists are scientists who study fossils to find out about life and climates of Earth long ago. Paleontologists work in many places, including dig sites, museums, and laboratories. Paleontology ties together many areas of science, including geology, botany, and biology. Someone who wishes to become a paleontologist should expect to take many different types of science classes and to be proficient in computer science.

Paleontologists collect fossils in the field. They bring them back to the laboratory for close study.

Astronomer

Do you enjoy stargazing and observing the patterns of the Sun and Moon? If you were an astronomer, studying these celestial bodies would be part of your job every day. Astronomers study everything from planets and moons to the formation of stars and galaxies. They use tools such as telescopes, infrared cameras, and spacecraft to gather data about space objects. Physics, astronomy, astrophysics, and math are some of the courses you should expect to take when studying astronomy.

Computers help astronomers make sense of data collected by radio telescopes, satellites, and spacecraft sent out through the solar system.

(t) Arpad Benedek/Getty Images; (b) Mark Sowa/NASA

Physical Science Careers

Materials Engineer

What material would be best for building a home in a hot climate? How can a bridge be made strong enough to withstand an earthquake? A materials engineer helps answer these questions and many more. By working with a variety of materials, including plastics, metals, and ceramics, a materials engineer determines which ones best fit certain needs. Students who want to become materials engineers need a degree in materials science or engineering. They also must have a strong background in math and physics.

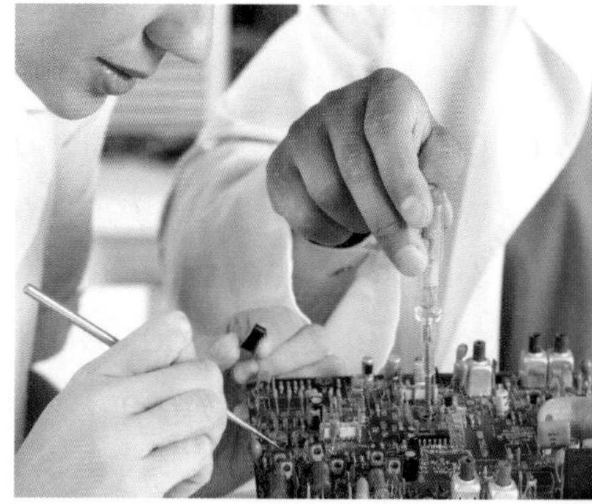

Materials engineers develop, make, and test materials for a wide range of products.

Construction Manager

If you have an interest in building, architecture, and design, then you may want to consider a career as a construction manager. Construction managers help plan and carry out a building project. They have knowledge of many different building materials and how those materials are affected by climate and weather. Construction managers also use math skills to help keep a project on budget. Some classes a construction manager might need include woodworking, engineering, architecture, math, and physics.

A construction manager coordinates all the activities on a building project.

Transportation Engineer

When you use a crosswalk or stop at a red light, you can thank a transportation engineer. Transportation engineers plan and design roadways, airport runways, highway systems, and railways that move people and goods from place to place. They study how traffic flows through intersections and determine where traffic lights and crosswalks should be placed. Most transportation engineers have a degree in civil engineering with a strong background in math and physical science.

Transportation engineers use computers to simulate traffic flow on streets, in crosswalks, or on train tracks.

Machinist

Many of the objects that you use every day, from your toothbrush to your bicycle, were made using a machine. Machinists build, use, and maintain these machines. When a machine is not working correctly, a machinist fixes it. Machinists also inspect the goods that machines make to be sure there are no problems or defects. Machinists have a strong math background, including geometry and trigonometry. They also take classes in blueprint reading, drafting, and computer science.

Machinists use lathes, milling machines, and grinders to produce precision metal parts. Some even use lasers to cut metal into precise shapes.

Wind Power Worker

Wind turbines are very large machines with many parts. It takes several people with different skills to manufacture, operate, and maintain them. Some wind power workers are engineers who design the turbines. Others are wind power technicians who work in the field to inspect the turbines and then fix them if something breaks. Several are scientists who conduct research or determine whether a place is suitable for using wind energy. Many of these workers have backgrounds in engineering, atmospheric science, and environmental science.

The growth of the wind energy industry will provide opportunities for a wide variety of careers.

Navigator

Imagine being on a ship in the ocean or on a jet plane flying 8 kilometers (5 miles) above Earth's surface. How does the captain know where to go so that the ship or plane can arrive safely? It is up to the navigator to determine that. Using the position of the Sun, Moon, and stars, as well as radar, magnetic compasses, and satellite data, a navigator helps guide a ship or plane safely from one point to another. Becoming a navigator requires training, as well as a background in computer sciences, math, physical science, geography, and oceanography.

Whether in a plane or in a boat, the navigator must know where the vessel is at all times.

Audiologist

If you enjoy helping people and have an interest in hearing and speech, then you might want to consider a career as an audiologist. Audiologists are medical professionals who diagnose and treat hearing and balance problems. They give hearing tests using a tool called an audiometer and, if necessary, help fit a patient with hearing aids. Becoming an audiologist takes approximately eight years of schooling. While in school, audiology students take courses involving the human body, physics, genetics, and computer science.

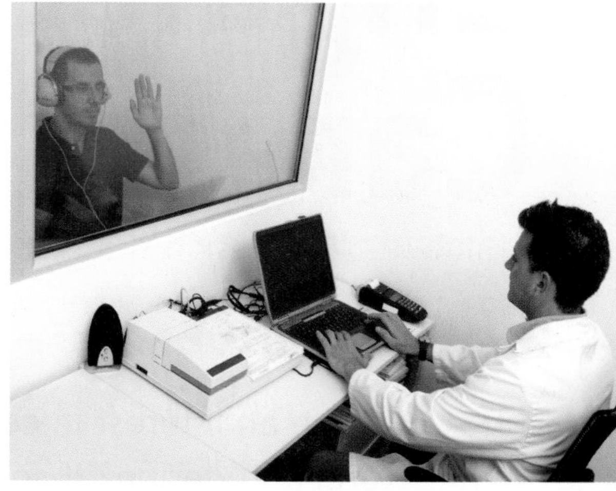

Audiologists use advanced technology and procedures to diagnose and treat hearing and balance problems.

Optician

If you were to borrow a friend's glasses, you might notice that they do not fit. That could be because an optician makes each pair of glasses to fit the dimensions of a particular person's face. Opticians use tools to take measurements, such as the distance between a person's pupils. Then they use those measurements to make glasses. Other tasks include fixing broken glasses and helping people choose the right contact lenses. Some opticians cut lenses and insert them into frames. Opticians are required to have a high school diploma and special on-the-job training.

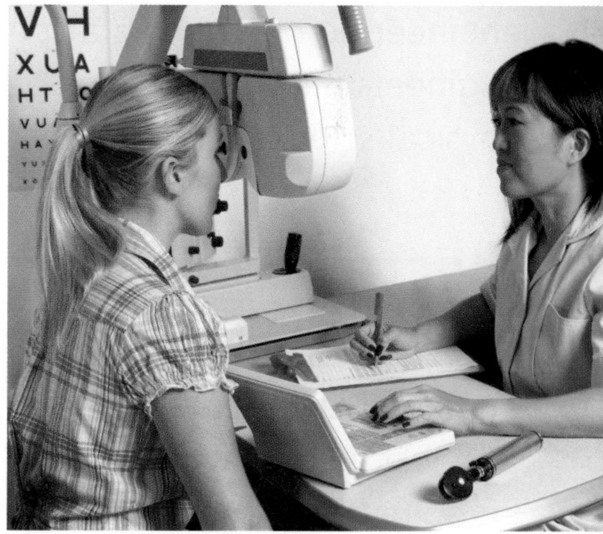

Opticians must be very precise. Eyeglasses that do not fit perfectly will not allow lenses to correct vision.

Engineering Careers

Civil Engineer

Do you have an interest in construction, architecture, and design? If so, you might want to consider a job as a civil engineer. Civil engineers plan, design, and oversee large-scale construction projects, including highways, tunnels, airports, dams, and bridges. They also oversee the construction of systems, such as those for distributing water or treating sewage. They work in an office or at a construction site. In addition to earning an engineering degree and a special license, civil engineers must also have a background in math and physical sciences.

Bridges are just one kind of large-scale project that civil engineers oversee.

Electrical Engineer

Next time you turn on a light, thank an electrical engineer! Electrical engineers plan, design, make, and test electrical equipment, which includes equipment that generates electrical power. Electrical engineers may also design communication systems, automobile engines, and radar equipment. They may work in an office in an engineering firm or in a factory. While earning an engineering degree, students study math, physics, drafting, and circuitry.

Electrical engineers often supervise the installation of complex electrical equipment.

Mechanical Engineer

Are you interested in knowing what makes the blades on a ceiling fan spin, or how the parts inside a car help it move? If an object has moving parts, then chances are a mechanical engineer was involved in designing it. From batteries to wind turbines, mechanical engineers work on a variety of projects. They must have a strong understanding of motion, forces, and math. They must also be creative and good at solving problems.

Mechanical engineers design and maintain mechanical equipment and parts.

Biomedical Engineer

The human body is much like a machine. Its parts must all work together in order for it to function. When one part isn't working correctly, a biomedical engineer might be able to fix it. As a biomedical engineer, you could design a part that helps a person's heart work better or build a new leg for someone who needs one. Some even work to help astronauts sleep better in space. Becoming a biomedical engineer requires at least four years of college, where you may study the human body, biology, communications, and computer science.

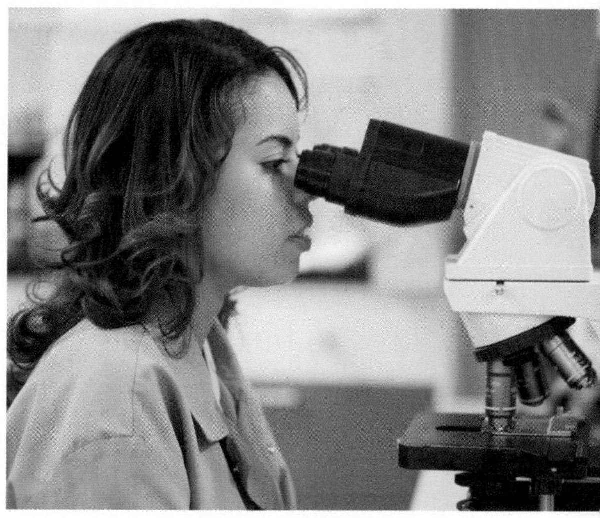

Biomedical engineers help people live healthy, active lives.

Chemical Engineer

If you enjoy learning about chemicals and are interested in improving peoples' lives, you may want to consider chemical engineering. Chemical engineers improve food-processing techniques. They construct the fibers that make clothes more comfortable. They develop ways to make life saving drugs more available and affordable. Chemical engineers work in factories, drug companies, electronic companies, plastics manufacturers, and biotechnology firms. In addition to chemistry, students studying chemical engineering take courses in biology, physics, math, computer science, and English.

Chemical engineers help devise ways of reducing air and water pollution and controlling toxic wastes.

Marine Energy Engineer

Oceans cover more than 70 percent of Earth's surface. Finding ways to harness energy available in and over oceans is the job of a marine energy engineer. These engineers design and build structures to harness the energy in waves and tides. In their work, marine energy engineers use computer models to test the effectiveness of their designs. They analyze the final designs to estimate the cost of a project. Some marine energy engineers oversee the installation of the system at sea. Students interested in becoming marine energy engineers usually major in structural or mechanical engineering. They also take courses in physics and oceanography.

Once installed in the ocean, generators like these tap the natural motion of ocean water and change it to usable energy.

Computer Hardware Engineer

Do you enjoy taking objects apart so that you can see how they work? If so, then you might enjoy a career as a computer hardware engineer. Hardware refers to the parts inside and outside a computer that make it work. Computer hardware engineers design and build these parts. They may work on small computers like the one you use at school or on huge servers that can store large amounts of data. Some work on computers that are found in appliances, cars, and smartphones. Once a computer is built, it is the hardware engineer's job to test it and then analyze the results. Computers are always changing, so hardware engineers must also make updates to computers as needed.

If you want to become a hardware engineer, you will need to receive a degree from a college or university. While in college, you will take classes in computer programming and other computer sciences. Hardware engineers must also have good critical thinking and problem solving skills.

Computers are in more and more of the objects that surround us. Computer hardware engineers make sure these computers function.

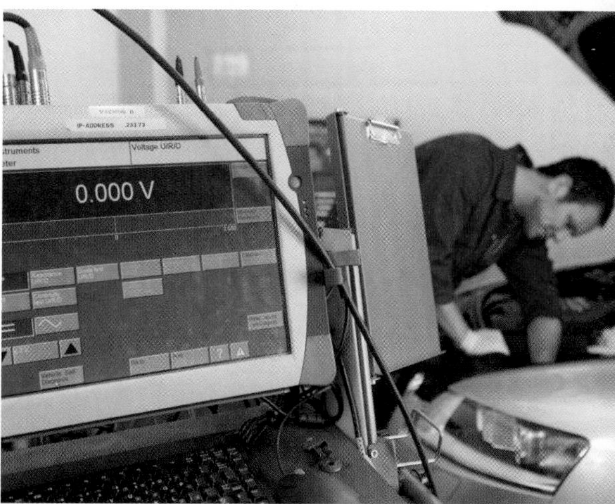

Modern cars have computers that sense emissions, tire pressure, and engine problems. Computer hardware engineers design not only those computers but also the computers that interpret the data so that a car can be repaired.

Overview **Science and Engineering Guide**

Science exploration uses many tools to measure data and to interpet information. Learn about them in this section!

Science Guide

Using Maps

A *map* is a drawing that shows an area from above. It shows a large area of Earth on a smaller, flat surface. Maps show where things are located. They show natural features like rivers, lakes, and mountains. They can also show features made by people, such as cities, roads, buildings, and parks.

Types of Maps

There are many different types of maps. Physical maps show Earth's land and water features. You could find the shape of an ocean or the name of a continent from a physical map.

Political maps show boundaries. You can tell how places are divided into countries, states, or cities on these maps.

Ocean and Continents

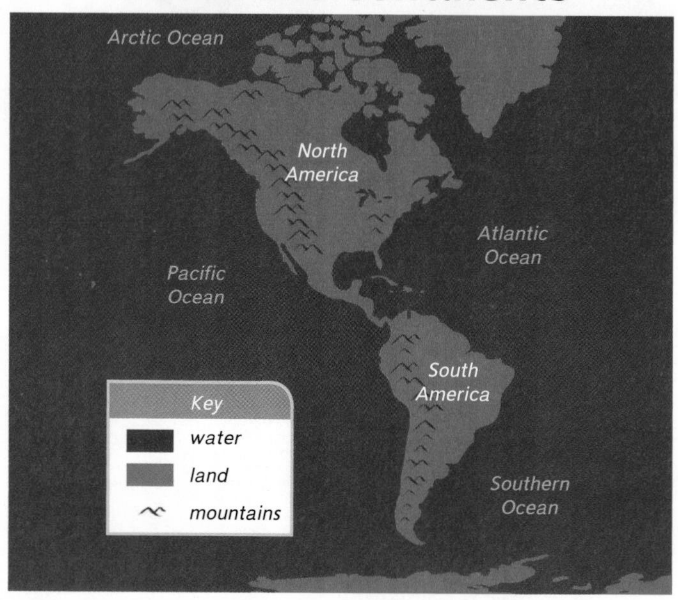

New England States

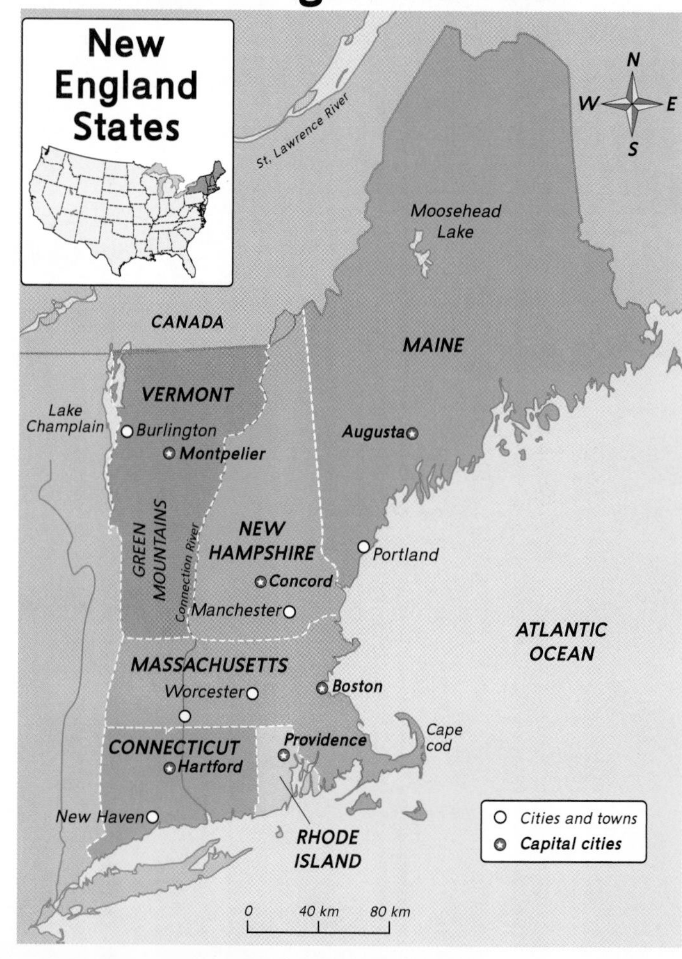

Road maps show roads and highways. Some show buildings and parks too. They help you find your way through towns, cities, and states.

Connecticut Highways

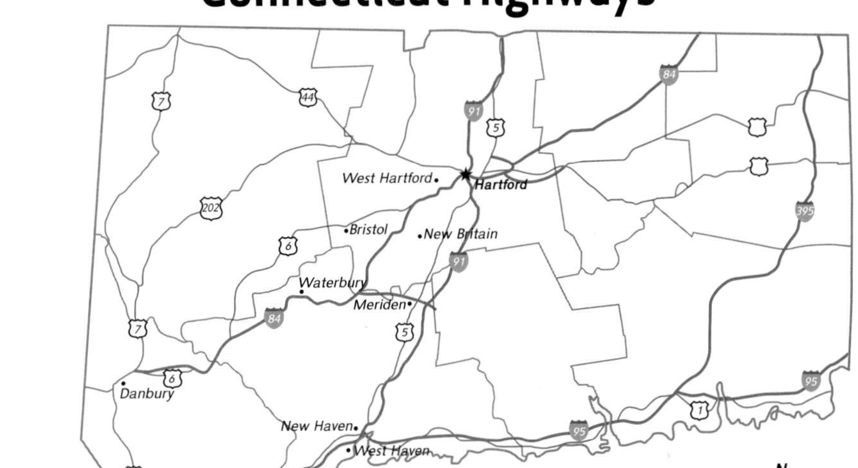

Some maps have special purposes. They might show climates of North America, world religions, or the rides in an amusement park. Some show how high or low the land surface is.

U.S. Climate Zones

CANADA
0 200 400 miles
0 200 400 kilometers
PACIFIC OCEAN
ATLANTIC OCEAN
MEXICO
Gulf of Mexico
ALASKA
HAWAII

tropical wet	semiarid	marine west coast	subarctic
tropical wet and dry	Mediterranean	humid continental cool	tundra
arid	humid subtropical	humid continental warm	highland

Reading Maps

Maps represent real places. To read a map, start with the title. It identifies what place on Earth the map represents. The title on the map below identifies the area shown as the city of Niagara Falls in New York.

Niagara Falls, New York

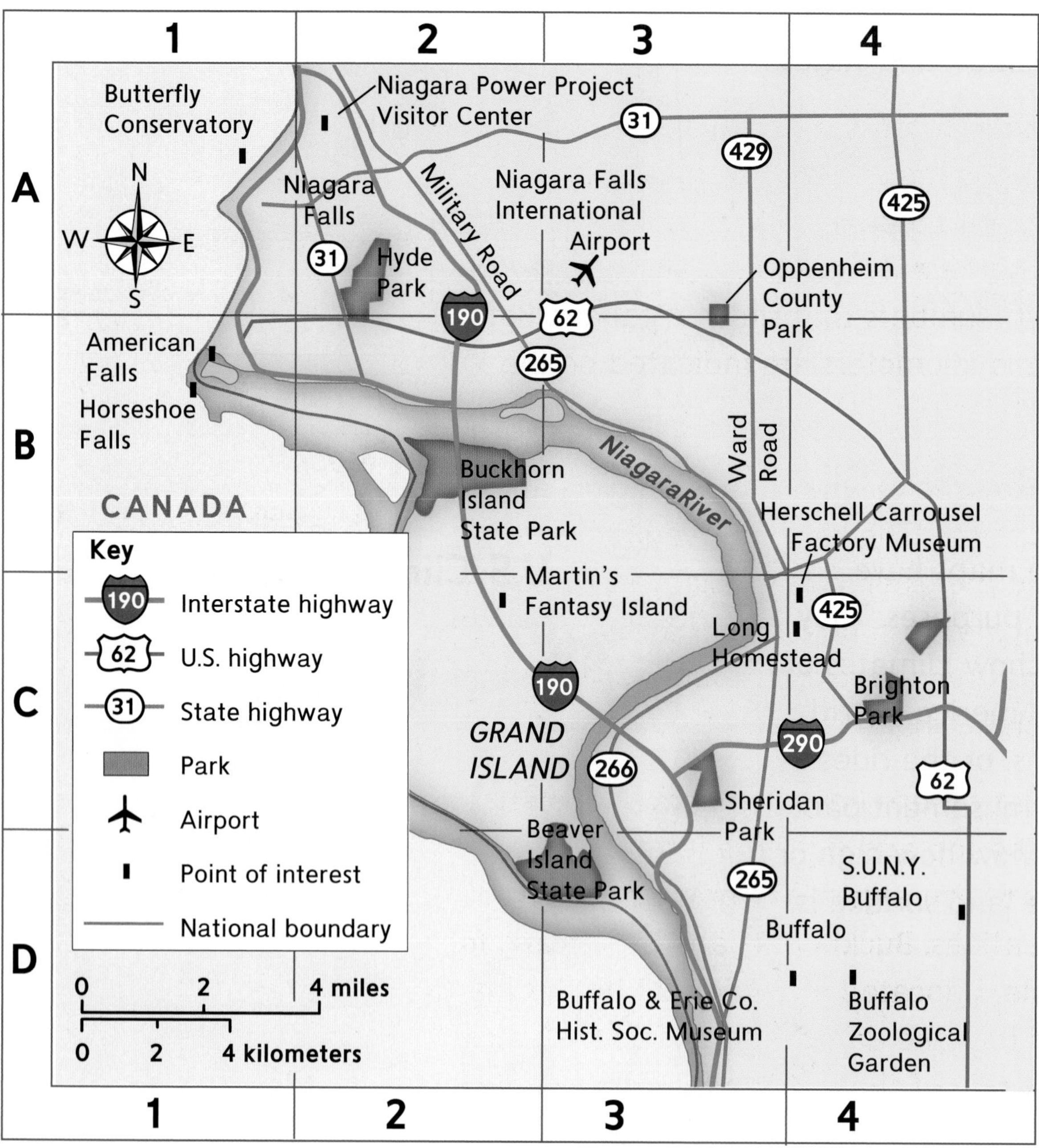

This map represents the city of Niagara Falls in New York.

Map Features

Compass Rose Maps also show where real things are located. That means they show directions. Directions are north (N), south (S), east (E), and west (W). Most maps show which way is north. Notice on the map shown here that north is up. Using that information, you can tell that Niagara Falls International Airport is east of Military Road.

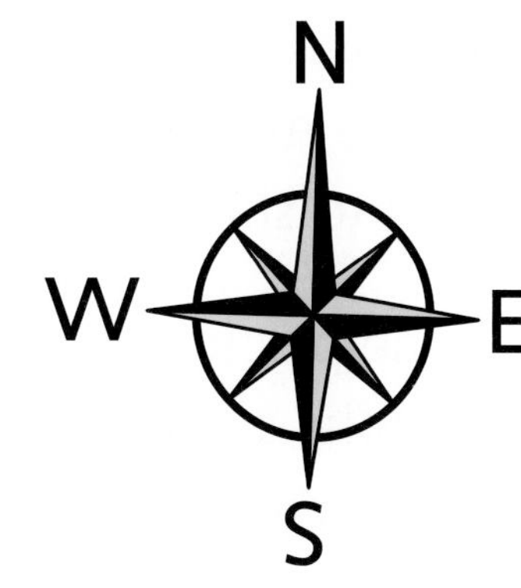

Map Scale To find out how far apart features are, use the map scale. The scale shows how distance on the map relates to distance on Earth's surface. This map's scale looks like a bar with numbers that show distances. Both miles and kilometers are indicated on the scale.

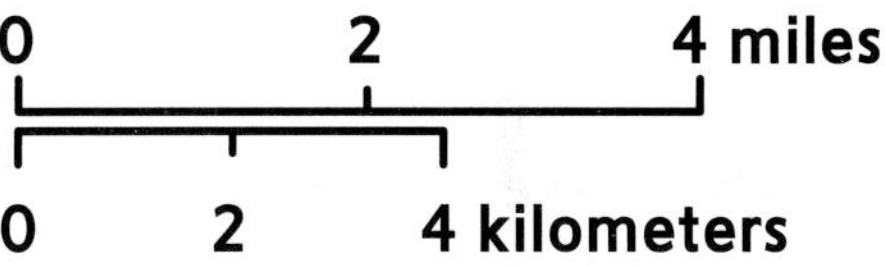

Key A key shows what a map's colors and shapes mean. The key helps you understand and find things on a map. Many maps use colored lines to show roads. This map key tells you that green areas represent parks and that an airplane shape represents an airport.

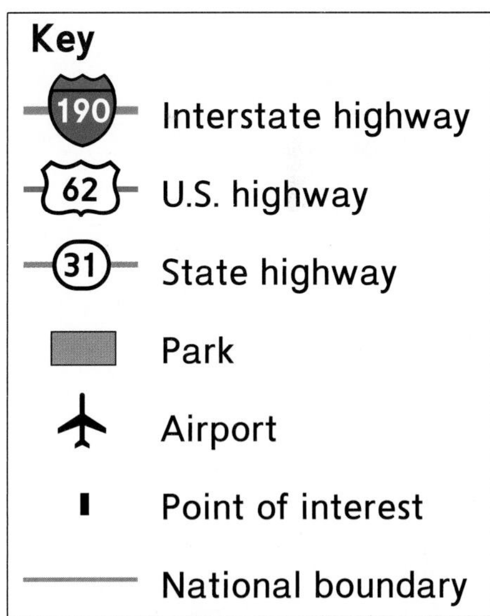

Grid Many maps have numbers and letters along the top and side. The letters and numbers form a grid. The grid can help you find locations. Buckhorn Island State Park, for example, is located in section B2 on this map. To find it, place a finger on the letter B along the left side of the map and another finger on the number 2 at the top. Move your fingers straight across and down until they meet.

How Scientists Use Maps

Many scientists depend on maps. They use maps to organize and analyze information. They also use maps to share information with others.

Maps help scientists understand living things. For example, scientists studying lizards make maps to show where the organisms are found. The maps help them compare where they predict lizards are likely to live and where they see them. Their map helps them find patterns in their data and understand what animals need to live.

Scientists use maps to show where different kinds of landforms, soils, and rocks are found, too. Studying these maps help them better understand Earth's features and how they change.

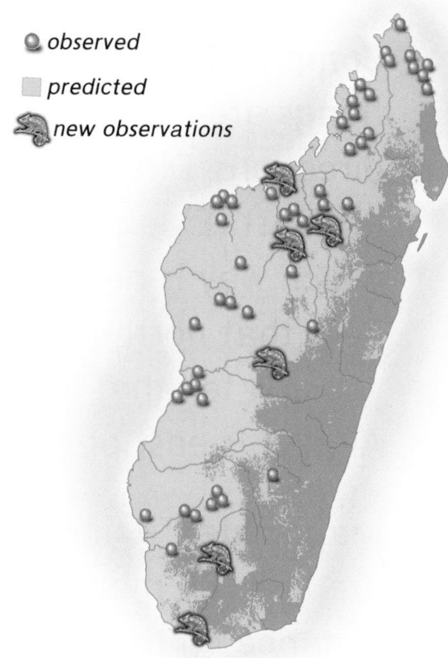

This map of Madagascar helps scientists understand lizards and their ecosystems.

This landform map uses color to show high and low features of Earth's surface. You can use it to find out where most mountains are found in the United States.

Weather scientists collect a lot of data. They organize the data they collect on weather maps. Weather maps show weather conditions for a certain area. They tell about air temperature, precipitation, clouds, and winds. They help scientists make predictions. They also help scientists share weather information with others who need it.

Weather maps help farmers decide when to plant and harvest their crops. They help pilots fly their planes safely and travelers plan their trips. They also help people decide what to wear and do each day.

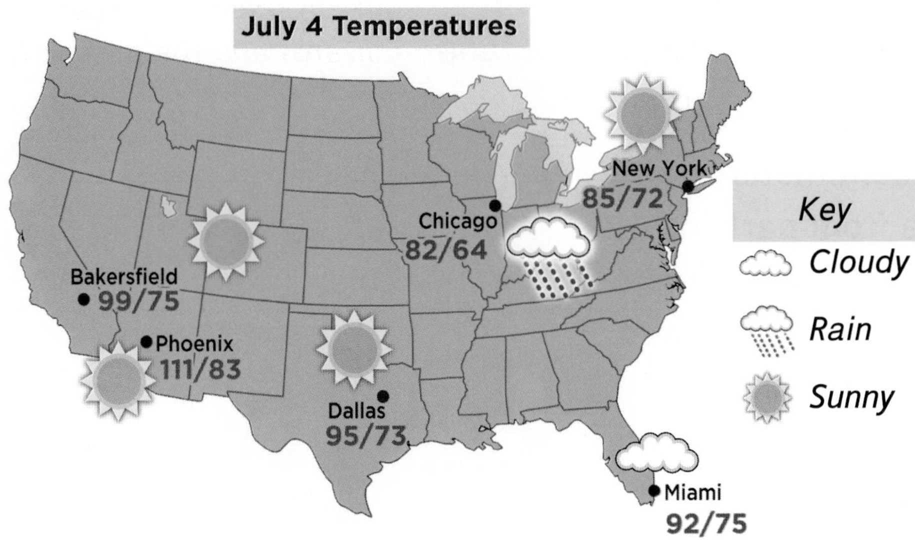

Weather maps help scientists organize weather data and make predictions.

Skill Builder **Read a Diagram**

Words are often included on maps. This weather map includes the names of major cities.

Using Math

Units of Measurement

Math is an important part of science. One way scientists use math is to measure things. They measure to find the length, volume, area, mass, weight, and temperature.

A measurement includes both a number and a unit. A unit of measurement that people agree to use is a standard unit. Scientists use a system of standard units called the metric system.

In the metric system, length and distance are measured in meters (m). Millimeters (mm) and centimeters (cm) are smaller units made from parts of meters. Kilometers (km) are larger units made of many meters. Speed measures the distance an object moves in a certain amount of time. So, the unit for speed is meters per second (m/s).

You can use the metric unit of meters per second to measure the speed that this girl moves.

Volume is often measured in units called liters (L). Sometimes volume is measured in cubic centimeters (cm^3) or milliliters (mL). Metric units for mass are grams (g) and kilograms (kg). Area is measured in square meters (m^2). The unit for weight is the newton (N). Temperature is measured in degrees Celsius (°C).

Another system of standard units used in the United States is the *customary system.* In this system, length and distance are measured in inches (in.), feet (ft), and miles (mi). Fluid ounces (fl oz) are used to measure liquid volume. Temperature is measured in degrees Fahrenheit (°F).

Table of Measurements

International System of Units (SI)	Tools	Customary Units
Temperature Water freezes at 0°C (degrees Celsius) and boils at 100°C.		**Temperature** Water freezes at 32°F (degrees Fahrenheit) and boils at 212°F.
Length and Distance 1,000 meters (m) = 1 kilometer (km) 100 centimeters (cm) = 1 meter (m) 10 millimeters (mm) = 1 centimeter (cm)		**Length and Distance** 5,280 feet (ft) = 1 mile (mi) 3 feet (ft) = 1 yard (yd) 12 inches (in.) = 1 foot (ft)
Volume 1,000 milliliters (mL) = 1 liter (L) 1 cubic centimeter (cm^3) = 1 milliliter (mL)		**Volume** 4 quarts (qt) = 1 gallon (gal) 2 pints (pt) = 1 quart (qt) 2 cups (c) = 1 pint (pt) 8 fluid ounces (oz) = 1 cup (c)
Mass 1,000 grams (g) = 1 kilogram (kg)		**Mass** 2,000 pounds (lb) = 1 ton (T) 16 ounces (oz) = 1 pound (lb)
Weight 1 kilogram (kg) weighs 9.81 newtons (N).		

(t-b) Burke/Triolo/Brand X Pictures/Jupiterimages; Amos Morgan/Getty Images; Polka Dot Images/Getty Images; Studiohio; Richard Hutchings

Changing Metric Units

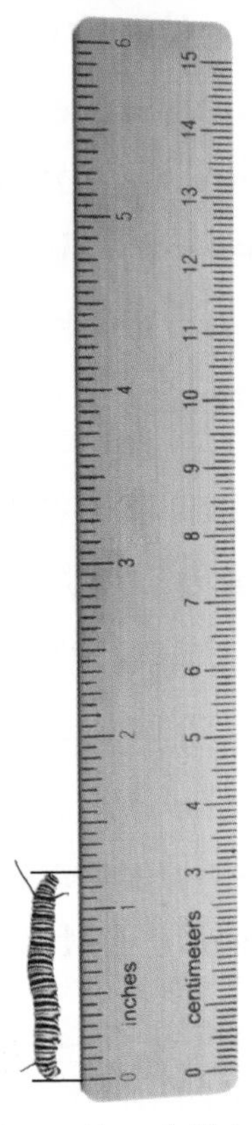

The length of this caterpillar is about 3 cm.

Suppose you measure the length of an object as 2 meters. But you need to share your results with the class in centimeters. In the metric system, it is easy to change a measurement from one unit to another. To change a metric unit, you multiply or divide by 10, 100, or 1,000.

Use the table below to find out how to change meters to centimeters. Find a row with "meters" in the first column and "centimeters" in the second column. Then look under "What to do" in that row. You should multiple by 100. Why? Because 1 m is equal to 100 cm.

2 meters x 100 = 200 centimeters

How to Change Metric Units of Length or Distance

Change	To	What to Do	Why
meters (m)	centimeters (cm)	multiply by 100	1 m = 100 cm
centimeters (cm)	meters (m)	divide by 100	
meters (m)	millimeters (mm)	multiply by 1,000	1 m = 1,000 mm
meters (m)	kilometers (km)	divide by 1,000	1,000 m = 1 km
centimeters (cm)	millimeters (mm)	multiply by 10	1 cm = 10mm
millimeters (mm)	centimeters (cm)	divide by 10	

Did You Know?

There are other ways to measure length other than meters and their related units. Within the solar system, astronomers use astronomical units (AUs) to measure distance. One AU is equal to the average distance from the center of the Sun to the center of Earth. Earth is 1 AU from the Sun. Neptune is about 30 AUs from the Sun.

To change units of other kinds of metric measurements, you also multiply or divide by 10, 100, or 1,000. The table below shows how to change units of volume and mass.

Suppose you find the mass of several rocks to be 2 kilograms, but your teacher wants your results in grams. You need to change the kilograms to grams. Look for a row in the table with "kilograms" in the first column and "grams" in the second column. Then look under "What to do" in that row. Because 1,000 g equals 1 kg, you should multiply the kilograms by 1,000.

2 kilograms x 1,000 = 2,000 grams

How to Change Metric Units of Volume and Mass

Change	To	What to Do	Why
liters (L) or grams (g)	milliliters (mL) or milligrams (mg)	multiply by 1,000	1 L = 1,000 mL 1 g = 1,000 mg
milliliters (mL) or milligrams (mg)	liters (L) or grams (g)	divide by 1,000	
grams (g)	kilograms (kg)	divide by 1,000	1,000 g = 1 kg
kilograms (kg)	grams (g)	multiply by 1,000	

Math Formulas Used in Science

You can use tools such as a ruler to measure length and a balance to measure mass. Other properties cannot be measured directly. Instead, different quantities must be added, subtracted, multiplied, or divided. In other words, these properties must be calculated using a formula.

Volume

To find the volume of an oddly shaped object, such as a rock, you must complete several steps. First, measure the volume of some water. Then place the object in the water and measure the volume again. Use these measurements and this formula to find the object's volume.

Volume of object = Volume of object and water – Volume of water

Volume of the rock =
860 mL - 550 mL = 310 mL

If you have an object shaped like a rectangle or cube, you can use the formula below to find the object's volume. First, measure how long, wide, and high the object is. Then multiply to find its volume.

Volume = length x width x height

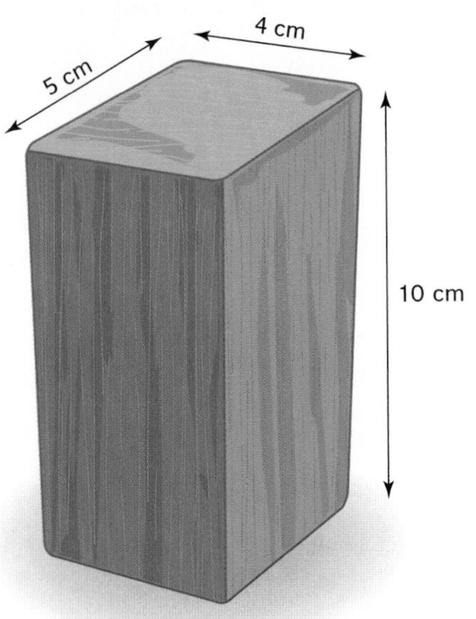

Volume = 5 cm x 4 cm x 10 cm = 200 cm^3

Ken Karp/McGraw-Hill Education

Area

You also multiply to find area. For a rectangle or square, like a tabletop, measure its length and width. Then use this formula to determine the area.

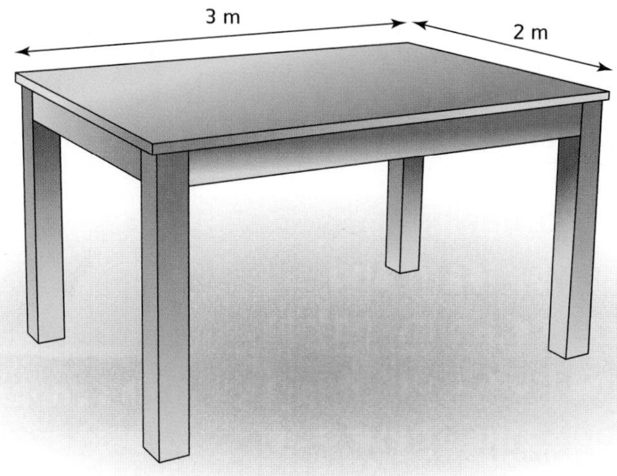

Area = 3 m x 2 m = 6 m^2

Speed

To find the speed of an object, you need to know two things. You need to know the distance the object traveled. You also need to know how much time it took to travel that distance. Then you divide the distance by the time, using the formula below.

Speed = *distance/time*

Suppose a student riding a skateboard traveled 100 meters in 50 seconds. You can use the formula to find his speed.

Speed = *distance/time* = 100 m/50 s = 2 m/s

If you measure distance and time, you can use a formula to find this boy's speed.

Using Language

Science Word Parts

When you study science, you learn many new words. These words may seem difficult to learn and remember at first. But there are steps you can take to help you. You can learn and look for common word parts used in science.

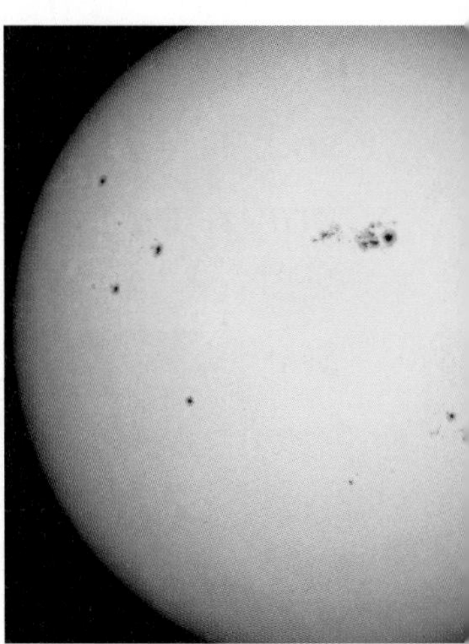

This photograph was taken with a special len[s]. It allows you to see the Sun's photosphere. This part of the Sun gives of[f] heat and light to Earth. Its name includes the root word photo, which means "light."

Like a plant, words have roots and other parts. Understanding these parts can help you learn and remember science words. Many science words did not start as words in English. They came from older words in other languages, such as Greek and Latin. These older words are called root words.

For example, the root word *thermos* is a Greek word meaning "hot." Knowing this root word may help you remember a word like *thermometer*, a tool that measures how hot something is. The table below lists science root words.

Science Root Words

Greek or Latin Root	What It Means	Example(s)
geo	Earth	geology
metron	measure	meter, kilometer, barometer, thermometer
photo	light	photosynthesis
planeta	to wander	planet, planetary
skopos	seeing	telescope, microscope
solaris	Sun	solar system, solar energy, solar eclipse
therme; thermos	heat; hot	thermal energy, thermometer, thermostat
Vulcan	Roman god of fire	volcano

A prefix is a word part added before a root word. It changes the meaning of the root word. Compare the words *telescope* and *microscope*. Both have the root word *scope* that comes from *skopos*, a Greek word for "seeing." Both the telescope and microscope are tools used for seeing things. But the prefix *tele-* means "far," so a telescope is a tool for seeing faraway objects. The prefix *micro-* means "very small," so a microscope is a tool for seeing very small objects.

telescope

Science Prefixes

Prefix	What It Means	Example(s)
tele-	far	telescope, television
micro-	very small	microscope
eco-	house/home	ecosystem, ecology
re-	again	recycle, resource, renewable, reflect
non-	not	nonrenewable
exo-	out of	exoskeleton
centi-	one hundredth or 1/100	centimeter
milli-	one thousandth or 1/1,000	millimeter, milliliter

microscope

A suffix is a word part added after a root word. It also changes the meaning of the root word. Think about the root word *geo,* which means "Earth." Add the suffix *-ology* and you get *geology*, the study of Earth. But if you use the suffix *-logist* instead, then you have *geologist,* a person who studies Earth.

Science Prefixes

Suffix	What It Means	Example(s)
-able	that can or will	renewable, nonrenewable
-ation	action or process	reaction, evaporation
-logist; -ist	a person who does an action	biologist, geologist, chemist
-ology	study of	biology, geology, ecology

Research Skills

When scientists ask questions, they conduct research to find answers. To research means to study and gather information about a topic. A modern way to conduct research is to search for information on the Internet.

Tips for Using the Internet

The Internet connects your computer with computers around the world, so you can use it to collect all kinds of information. But not all information on the Internet is correct. And not all Web sites are safe.

When using the Internet, visit only Web sites that are safe and reliable. Reliable means trustworthy. It is based on facts, not on what people believe or feel. Science facts are proven by observations and data. Your teacher can help you find safe and reliable sites to use.

You can use the Internet to gather information about many different science topics.

damircudic/Getty Images

Look for Web sites with information written by real scientists. They include Web sites run by the U.S. and state governments. You might also try sites run by colleges, museums, and professional science groups.

New science data are collected every day. Make sure you use the latest information possible. Information that is old may no longer be correct.

Check the information on more than one reliable Web site. If you find the same information on both a museum and a university site, it is more likely to be factual.

Avoid Web sites that try to sell products or ask you for personal information. These sites may not be reliable or safe.

Did You Know?

The Internet started in 1969 with a network of just four supercomputers. Today that network is made up of more than 30 million users.

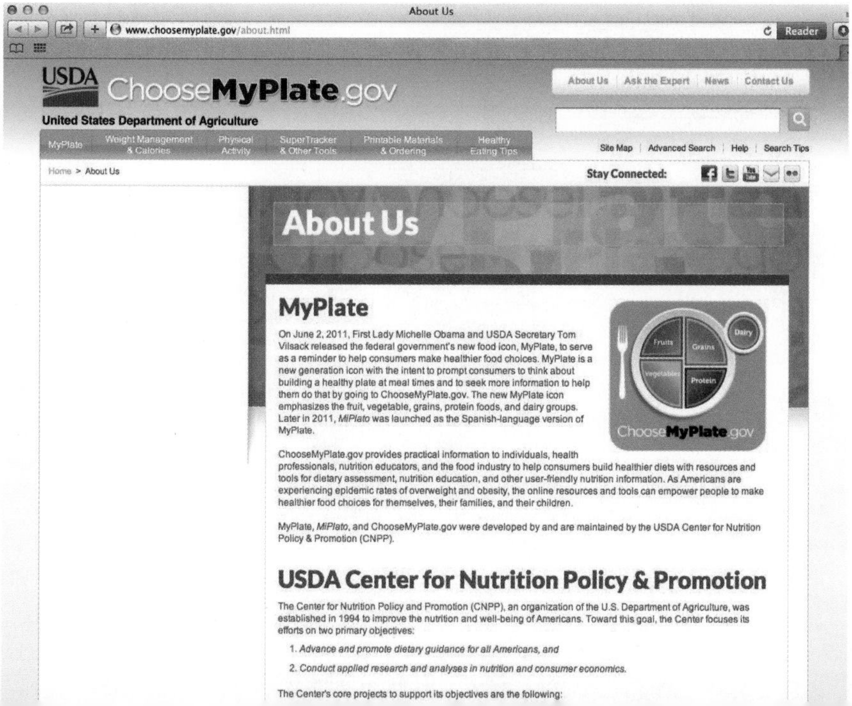

USDA's Center for Nutrition Policy and Promotion

Science information on government Web sites, such as the U.S. Department of Agriculture site, is written by real scientists. That means the information is more likely to be reliable.

Finding the Right Materials

The Internet is only one of many different sources of science information. You can also do research with materials in your school media center or library.

Use your topic or key words to search in your library catalog. Look for books and articles written by scientists. You might find articles in newspapers. You might find them in science magazines called journals. Books and magazines have a table of contents at the front and an index at the back. These sections can help you find your topic quickly.

Research materials may be on paper. They might be digital, such as e-books, online newspapers, compact discs (CDs), and digital video discs (DVDs). No matter what kind of materials you have, always use the newest materials in your media center or library.

Your school media center or library has many materials you can use to do science research.

If you cannot find the information you need, don't panic. Ask a teacher or librarian for help. They can suggest other research materials to use.

You might even try asking a scientist or other expert what you want to know. Ask your teacher to help you find and contact a scientist. An expert on your research topic might live or work in your community. Your teacher could also help you send your questions by e-mail or in a letter to a scientist who lives far away.

Take your time, ask for help when you need it, and keep trying. Soon you will have the skills to find the right research material and get the science information you need.

A scientist in your community may be able to answer your research questions.

Steve Debenport/iStock/Getty Images

Glossary

Pronunciation Key

The following symbols are used throughout this glossary.

a	at	e	end	o	soft	u	up	hw	**white**	ə	**a**bout
ā	ape	ē	me	ō	go	ū	use	ng	so**ng**		tak**e**n
ä	farther	i	tip	ôr	fork	ü	rule	th	**th**in		penc**i**l
âr	care	ī	ice	oi	oil	ù	pull	t͟h	**th**is		lem**o**n
ô	law	îr	fear	ou	out	ûr	turn	yü	**ru**le		circ**u**s
								zh	mea**s**ure		

ˊ= primary accent; shows which syllable takes the main stress, such as **at** in **atmosphere** (atˊmə·sfîrˊ)

ˊ = secondary accent; shows which syllables take lighter stresses, such as **sphere** in **atmosphere**

accelerate ak·seˊlə·rātˊ
a change in velocity with respect to time

accommodation ə·käˊmə·dāˊshən
an individual organismˊs response to change in its ecosystem

adaptation adəp·tāˊshən
a structure or behavior that helps an organism survive in its environment

air mass âr mas
a large region of the atmosphere in which the air has similar properties

air pressure âr preˊshər
the weight of air pressing down on Earth

amber amˊbər
hardened tree sap, yellow to brown in color, that can be a source of insect fossils

atmosphere atˊmə·sfîr
the layers of gases that surround Earth

attract ə·traktˊ
a large, sudden movement of ice and snow down a hill or mountain

axis akˊsəs
a line through the center of a spinning object

balanced forces ba´lənst fôrs´əz
forces that act together on an object without changing its motion

behavior bi´hā·vyər
the way a person or animal acts or behaves

birth bərth
the beginning or origin of a plant or animal

blizzard bli´zərd
a storm with lots of snow, cold temperatures, and strong winds

camouflage kam´ə·fläzh´
an adaptation that allows an organism to blend into its surrounding

cast kast
a fossil formed or shaped within a mold

climate klī´mət
the pattern of weather at a certain place over a long period of time

cloud kloud
a collection of tiny water drops or ice crystals in air

competition kom´pə·ti´shən
the struggle among organisms for water, food, or other needs

compound machine kom´pound mə·shēn´
two or more simple machines put together

consumer kən·sü´mər
an organism that eats plants or other animals

decomposer dē´kəm·pō´zər
an organism that breaks down dead plant and animal material

direction də´rek·shən
the course or path on which something is moving

distance dis´təns
the amount of space between two objects or places

earthquake ûrth´kwāk
a sudden movement of the rocks that make up Earth´s crust

ecosystem ē´kō·sis´təm
the living and nonliving things that interact in an environment

electrical charge i·lek´tri·kəl chärj
the property of matter that causes electricity

endangered en·dān´jərd
when one kind of organism has very few of its kind left on Earth

energy e´nər·jē
the ability to do work

environment en·vī´rən·mənt
all the living and nonliving things that surround an organism

extinct ek·stingkt´
when there are no more of an organism´s kind left on Earth

flood flud
when dry land becomes covered with water

force fôrs
a push or a pull

fossil fä´səl
the trace of remains of something that lived long ago

friction frik´shən
a force that opposes the motion of one object moving past another

group grüp
a number of living things having some natural relationship

growth grōth
the process of growing in size or amount

hibernation hī´bər·nā´shən
to rest or go into a deep sleep through the cold winter

hurricane hûr´i·kān´
a large storm with strong winds and heavy rain

inclined plane in·klīnd´ plān
a simple machine with a flat slanted surface that is raised at one end

inherit in·her´ət
a characteristic that is passed from parents to offspring

instinct in´stingt´
an inherited behavior, one that is not learned but is done automatically

learned behavior lûrnd bi·hā´vyûr
behavior that is learned from watching others

lever le´vər
a simple machine made of a bar that turns around a point

life cycle līf sī´kəl
how a certain kind of organism grows and reproduces

load lōd
the object being moved by a machine

magnet mag´nət
an object with a magnetic force; magnets can attract or repel certain metals

magnetic field mag·ne´tik fēld
a region of magnetic force around a magnet, represented by lines

magnetism mag´nə·ti´zəm
the ability of an object to push or pull on another object that has the magnetic property

metamorphosis me´tə·môr´fə·səs
a series of changes in which an organism´s body changes form

migrate mī´grāt
to move from one place to another

migration mī´grat´shən
to move from one place to another

mimicry mi´mi·krē
an adaptation in which one kind of organism looks like another kind in color and shape

motion mō´shən
a change in the position of an object

motor mō´tər
a machine that causes motion or power

N

natural hazard na´cha ·rəl ha´zərd
a natural event such as flood, earthquake, or hurricane that causes great damage

natural selection na´chə·rəl sə·lek´shən
the natural process in which only the organisms that adapt to their environment survive and reproduce

nocturnal nok·tûr´nəl
an adaptation in which an animal is active during the night and asleep during the day

O

offspring of´spring
the child or young of a particular human, animal, or plant

organism ôr·gə·ni´zəm
a living thing

P

pole pōl
one of two ends of a magnet, where the magnetic force is strongest

pollution pə·lü´shən
what happens when harmful materials get into water, air, or land

population po·pyə·lā´shən
all the members of one kind of type of organism in an ecosystem

position pə·zi´shən
the location of an object

precipitation pri·si´pə·tā´shən
water that falls to the ground from the atmosphere

producer prə·dü´sər
an organism, such as a plant, that makes its own food

R

repel ri·pel´
to push away

reproduction rē·prə·duk´shən´
the making of offspring

season sē´zən
time of the year with different weather patterns

simple machine sim´pəl mə·shēn´
a machine with few or no moving parts

speed spēd
how fast an object moves over a certain distance

static electricity sta´tik i·lek´tri´sə·tē
the build up of an electrical charge on a material

tornado tôr·nā´dō
a powerful storm with rotating winds that forms over land

trait trāt
a feature of a living thing

unbalanced force un·ba´lənst fôrs´əz
forces that do not cancel each other out and that cause an object to change its motion

variation ver·ē´ā·shən
a difference among the same type of plant or animal

velocity və·lo´sə·tē
the speed and the direction of a moving object

weather we´thər
what the air is like at a certain time and place

wheel and axel hwēl and ak´səl
a simple machine made of a wheel stuck to a rod

wind wind
moving air

work wûrk
what is done when a force changes an object´s motion

Index

D

F

G

H

I

J

K

L

M

N

S

W